277
Current Topics in Microbiology and Immunology

Springer
Berlin
Heidelberg
New York
Barcelona
Hong Kong
London
Milan
Paris
Tokyo

A. RETHWILM (Ed.)

Foamy Viruses

With 40 Figures and 7 Tables

Springer

Prof. Dr. Axel Rethwilm

Medizinische Fakultät der TU Dresden
Institut für Virologie im MTZ
Fiedlerstr. 42
01307 Dresden
Germany
e-mail: axel.rethwilm@mailbox.tu-dresden.de

Cover Illustration by A. Rethwilm (this volume):
Foamy virus cytopathic effect in a VeroA cell culture infected with a simian foamy virus

ISSN 0070-217X
ISBN 3-540-44388-6
Springer-Verlag Berlin Heidelberg New York

Library of Congress Catalog Card Number 72-152360

Springer-Verlag Berlin Heidelberg New York
a member of BertelsmannSpringer Science+Business Media GmbH

http://www.springer.de

Springer-Verlag is a company in the specialist publishing group BertelsmannSpringer

Printed in Germany

Typesetting: Stürtz AG, Würzburg
Cover Design: Design & Production GmbH, Heidelberg
Production Editor: Angélique Gcouta, Berlin
Printed on acid-free paper SPIN: 10768969 27/3020 5 4 3 2 1 0

Preface

In the year 2002 the nomenclature of the *Retroviridae* was changed by the International Committee on Taxonomy of Viruses (ICTV). The previous arrangement of seven genera in one family subsequently became six genera within the subfamily of the *Orthoretrovirinae* and one genus (foamy viruses) in the subfamily *Spumaretrovirinae* (see Table 1).

There were a variety of reasons why the subcommittee in question suggested to the ICTV that the family of *Retroviridae* should be separated into two subfamilies. The main reason, however, was the aberrant replication pathway of foamy viruses, which distinguishes them from all other retroviruses.

Although we cannot present a complete picture, so many details have been accumulated over the past ten or so years, that it seems worthwhile to summarize them here. The first five chapters of this book deal with one or the other aspect of this peculiar replication strategy. Basic research will probably uncover further fascinating findings on the molecular biology of foamy viruses. However, due to recent advancements, the application of foamy virus-derived vectors may dominate the interest in the future. Therefore, a comprehensive chapter on this aspect by Mergia and Heinkelein is also included.

The use of cell cultures and subgenomic plasmids dominated the research on foamy viruses in the past. However, a complete understanding of the foamy virus replication pathway requires animal experiments. The current knowledge on the replication of foamy viruses *in vivo* is reviewed by Falcone et al. in chapter seven.

Foamy viruses are naturally found in those species that harbour lentiviruses, with the exception of humans (see below). Recently a molecular characterization of viral isolates from bovines, felines, and equines has been undertaken. These results are summarized in chapter nine by Ali Sa.

However, still most of the findings on foamy viruses were obtained by research on a single virus isolate previously called "human foamy virus", which appeared to be the first to be investigated on a molecular level. However, to my knowledge genuine human foamy viruses do not exist. What does exist are several trans-species transmissions of different simian foamy viruses from monkeys and apes to human hosts. In chapter eight by

Heneine et al. these zoonoses are dealt with. Thus, the term "human foamy virus" is very much misleading. Therefore and for the sake of conformity in this book, the authors agreed to use the more neutral term "PFV" for prototypic foamy virus.

From the first plan until realisation this book had a very long incubation period. The reason for this is simply due to the fact that worldwide only a handful research labs continuously pursue work on foamy viruses. This obviously limits the number of potential authors. In particular, if one wants to be as up to date as we wanted to be with these reviews. However, if at least one reader becomes motivated to enter the field, the aim of this book will have been achieved.

Dresden 2003 AXEL RETHWILM

Table 1: The ICTV nomenclature of the Retroviridae

Subfamily	Genus	Example	Virion Morphology	Genome
Orthoretro-virinae	**Alpharetro-viruses**	Avian Leukosis Virus (ALV)	central, spherical core "C-type particles"	simple
	Betaretro-viruses	Mouse Mammary Tumor Virus (MMTV), Mason-Pfizer Monkey Virus (MPMV)	eccentric, spherical core "B-type particles" (MMTV) or cylindrical core "D-type particles" (MPMV)	simple
	Gammaretro-viruses	Murine Leukemia Virus (MuLV)	central, spherical core "C-type particles"	simple
	Deltaretro-viruses	Bovine Leukemia Virus (BLV), Human T-cell Leukemia Virus (HTLV)	central, spherical core "C-type particles"	complex
	Epsilonretro-viruses	Walleye Dermal Sarcoma Virus (WDSV)	central, spherical core "C-type particles"	complex
	Lentiviruses	Human Immuno-deficiency Virus (HIV)	cone-shaped core	complex
Spumaretro-virinae	**Foamyviruses**	Prototype Foamy Virus (PFV)	central, spherical core	complex

List of Contents

List of Contributors

(Their addresses can be found at the beginning of their respective chapters.)

Chapter 1

The Replication Strategy of Foamy Viruses

A. Rethwilm

Abstract The replication strategy of foamy viruses diverges in many aspects from what is commonly accepted as the rules of retroviral replication. Although many questions on the details of the replication pathway are still unanswered, it appears that foamy viruses have adopted a strategy which functionally bridges the retroviral and the hepadnaviral replication pathways. A number of experimental findings in favour of the view that foamy viruses are reverse transcribing DNA viruses which integrate into the host cell genome are discussed.

A. Rethwilm (✉)
Institut für Virologie, Medizinische Fakultät "Carl Gustav Carus",
Technische Universität Dresden,
Fetscherstr. 74, Dresden, 01307 Germany
E-mail: Axel.Rethwilm@mailboxtu-dresden.de

1
Introduction

At first sight foamy virus (FV) genomes look like ordinary retrovirus genomes: the provirus is flanked by long terminal repeats (LTRs), which can be subdivided into U3, R and U5 regions (Vogt 1997b). In between are located the canonical genes *gag*, encoding the capsid protein, *pol* for the enzymatic functions and *env*, specifying the envelope glycoprotein(s) (Vogt 1997b). In addition, there are at least two accessory open reading frames in the 3' genomic region, which give rise to two proteins of proven (Tas) or suspected (Bet) regulatory function (Fig. 1). Regulatory gene products are already known from other complex retroviruses (Vogt 1997b). In other words, at first sight FV genomes do not reveal spectacular new retroviral characteristics. This fact and the fact that we are dealing with a group of viruses with no known pathogenicity raises questions about the rationale for studying FVs.

However, and fortunately for the field, analysis of the details reveals a different story (Linial 1999). Whatever aspect of the virus is considered, be it the regulation of viral transcription, the structure and function of any of the viral proteins, or indeed any aspect of the FV replication cycle, extensive deviations from the generally accepted rules in retrovirology will be found. The reader is referred to the specialised chapters that follow dealing with these topics.

Although far from a complete understanding of the evolutionary advantages FVs must have gained, this chapter attempts to address the basic aspects of the FV replication strategy. Because many questions remain open, this summary is necessarily speculative in some aspects.

2
Reverse Transcription

The conversion of the retroviral RNA genome into linear double-stranded DNA is a highly ordered process conserved among all *Retroviridae* (Telesnitsky and Goff 1997). There is, so far, no indication that the principal order of events which lead to cDNA synthesis is fundamentally different between *Spuma-* and *Orthoretrovirinae*. The packaged RNA (pre-)genome (see below) is probably diploid, because FV RNA harbours at its 5' end dimerisation signals of functional importance (Cain et al. 2001; Erlwein et al. 1997). A $tRNA^{lys1,2}$ serves as primer for the synthesis

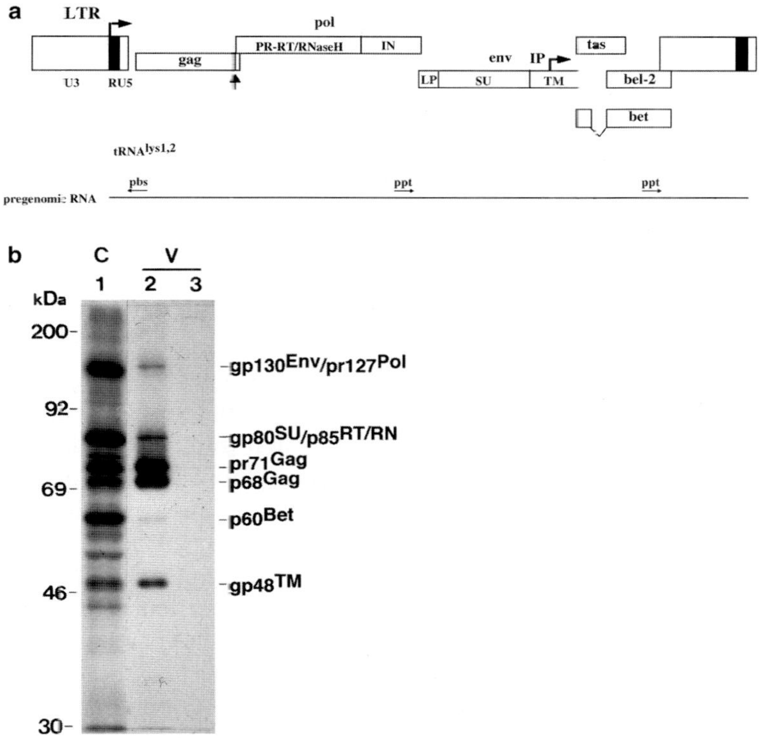

Fig. 1a, b. FV genome structure and main cellular and virion proteins. **a** FV DNA genome and open reading frames are displayed in the upper panel. The *vertical arrows* indicate the start of transcription from the two promoters which are located in the long terminal repeat (*LTR*) and the internal promoter located (*IP*) in the env gene. The *vertical arrow* indicates the approximate protease cleavage site in the C-terminal Gag protein. At the *bottom* the pre-genomic RNA with primer binding site (*pbs*) homologous to the cellular $tRNA^{lys1,2}$ and the two poly purine tracts (*ppt*) as start sites for plus-stranded cDNA synthesis are shown. **b** Main FV proteins as detected by immunoprecipitation (with serum from an infected animal) of radioactively labelled proteins from FV infected cells (lane *1*) or from virions (lane *2*). Lane *3* is a negative control lane. The two Gag protein species of 68 and 71 kDa apparent molecular weight dominate in extracellular virus, the 60 kDa Bet is one of the major intracellular proteins. The glycoproteins of 130 and 80 kDa apparent MW can usually only be distinguished from the Pol precursor protein of 127 kDa and the RT/RNaseH of 85 kDa when protein-specific antisera are used. (Modified from Netzer et al. 1990)

of minus-stranded strong stop cDNA (Maurer et al. 1988) (Fig. 1). As in some orthoretroviruses and, in particular, lentiviruses, there is a centrally located poly purine tract (*ppt*) which functions as a second start of transcription for plus-stranded cDNA in addition to the *ppt* located upstream of the U3 region (Kupiec et al. 1988; Telesnitsky and Goff 1997; Tobaly-Tapiero et al. 1991; Whitwam et al. 2001) (Fig. 1). The central *ppt* and adjacent sequences are required for efficient human immunodeficiency virus (HIV) replication (Charneau and Clavel 1991; Charneau et al. 1992; Charneau et al. 1994). It is likely that FVs profit in a similar way from a central *ppt*. However, formal experimental proof of this hypothesis does not exist.

Reverse transcription of orthoretroviruses occurs early in infection after penetration of a cell and disassembly of the viral particle (Vogt 1997a). Early products of reverse transcription (minus-stranded strong stop cDNA) have been detected in virions produced by HIV- or murine leukaemia virus (MLV)-infected cells by sensitive methods, but this appears to be the exception rather than the rule (Lori et al. 1992; Trono 1992). Functional studies with the reverse transcriptase (RT) inhibitor azido-thymidine (AZT) strongly suggest that inhibition of the generation of these reverse transcripts in virus-producing cells has no influence on the infectivity of orthoretroviruses, whereas its addition to newly infected cell cultures abolishes viral infectivity (Roy et al. 2003; Trono 1992). In other words, these studies confirm that orthoretroviruses are essentially RNA-containing viruses replicating through a DNA intermediate with the reverse transcription step being early in the replication cycle immediately following infection of a cell (Vogt 1997a). On the other hand, similar studies revealed that FVs behaved quite differently (Fig. 2). Although AZT had no significant influence when added to cells freshly infected with FVs, it completely blocked infectivity when added to the virus-producing cells (Moebes et al. 1997; Roy et al. 2003; Yu et al. 1999). This holds true for different FVs, when different cell types are used for virus production and infection, for virus produced by transient plasmid transfection, i.e. under synchronised conditions, or from persistently infected cultures (Moebes et al. 1997; Roy et al. 2003; Yu et al. 1999). All available evidence from the use of AZT as an inhibitor suggests that in FVs reverse transcription occurs late in replication and, therefore, at a very different point in the life cycle compared with orthoretroviruses (Fig. 2). It has long been noted that FV-infected cells may contain up to 1,000 or more copies of unintegrated linear viral DNA

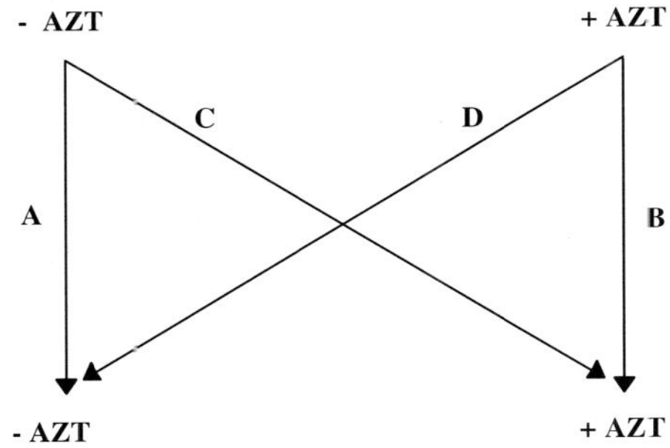

	A	B	C	D
Orthoretroviruses	100%	0-1%	0-1%	90-100%
Spumaretroviruses	100%	0-1%	90-100%	0-1%

Fig. 2. AZT inhibition experiment to clarify the time of reverse transcription in the FV replication cycle compared with an orthoretrovirus. In experiment *A*, virus was produced by transfection and the supernatant was titrated in the absence of the RT inhibitor AZT. In experiment *B*, virus production and titration were done in the presence of AZT. The important experiments are *C* and *D*, in which virus was either titrated or produced in the presence of AZT, respectively. The different results obtained with orthoretroviruses and FV are displayed at the *bottom*. (Modified from Moebes et al. 1997; Roy et al. 2003)

(Mergia and Luciw 1991; Moebes et al. 1997). In latently infected cells harbouring a single provirus this integrated DNA can no longer be detected by conventional methods once the active replication cycle has been induced, because the signal is too weak to be seen above the background of the many linear viral cDNA molecules (Schweizer et al. 1993). Late reverse transcription of FV genomes is the reason for this observation. In other words, FVs appear to be essentially DNA viruses replicating through an RNA intermediate. This feature is in sharp contrast to orthoretroviruses.

The viral molecules involved in reverse transcription of orthoretroviruses include, at least, the diploid RNA genome, Pol enzymes and the nucleocapsid protein (NC), a particular cleavage product of the Gag precursor with RNA binding and nucleic acid chaperone properties (Guo et al. 2002; Telesnitsky and Goff 1997). The NC is required for efficient minus-stranded strong stop cDNA synthesis and strand transfer reactions which lead to the completion of minus-stranded and plus-stranded strong stop cDNA synthesis (Guo et al. 2002; Telesnitsky and Goff 1997).

FVs do not have a NC protein. Moreover, except for a single C-terminal cleavage, which chops off a small peptide, further Gag cleavage products are absent from the intracellularly assembled capsids and are hard to detect in extracellular virions (see Fig. 1B; see also chapters in this book by Flügel and Pfrepper and by Linial and Eastman for a detailed discussion of this matter). Further Gag cleavages may occur on capsid disassembly very early in the replication cycle after uptake of virions by cells. In any case, such cleavage products are highly unusual (Pfrepper et al. 1999). Because cDNA synthesis takes already place in cells producing viruses, the conditions of FV reverse transcription are very different from those for orthoretroviruses. Unfortunately, much more cannot be said about these conditions. RNA, the *pol* reading frame proteins and the two Gag species (the uncleaved precursor and the slightly shorter N-terminal cleavage product) constitute the FV capsid, as far as the viral components are concerned. It is known that Gag protein and the C-terminal Gag cleavage are required for cDNA synthesis and virus infectivity, because a *gag* mutant unable to perform the cleavage was also found to be unable to generate viral DNA (Enssle et al. 1997). This may be taken as very indirect evidence that the functionally active FV genome must be DNA. However, because other effects of the mentioned *gag* mutation, such as a reduced RNA incorporation, were not formally excluded, this suggestion is highly speculative. The exact circumstances of FV reverse transcription, which include the identification of activating factors or other conditions required for cDNA synthesis, have not yet been resolved.

3
The Nature of the Foamy Virus Genome

Given the clear-cut evidence for a functionally relevant FV DNA genome that was obtained in the AZT experiments mentioned above, it is not surprising that linear double-stranded virion DNA has been physically detected by Southern blotting (Roy et al. 2003; Yu et al. 1996a). Furthermore, cells can be transfected with DNA extracted from purified extracellular virions, which leads to new rounds of infection (Roy et al. 2003; Yu et al. 1999). The infectious nature of the virion genome is the most salient difference between *Spuma-* and *Orthoretroviridae.*

The virion DNA probably is gapped on the plus strand around the location of the central *ppt* (Kupiec et al. 1988; Tobaly-Tapiero et al. 1991).

It is not yet known whether there are further areas in the genome with single-stranded DNA gaps, or whether there are molecules with stretches of viral RNA still attached, which have not been reverse-transcribed into minus-stranded DNA. In this case, the genome would be made up of a mixture of RNA and DNA. However, the AZT experiments argue against this being of functional relevance. Although apparently full-length linear DNA can be detected in Southern blots, researchers failed to detect genomic length RNA by Northern blotting (Yu et al. 1996a).

Anecdotally, it is worth mentioning that when the first molecular clones of a FV genome were generated, the probes to detect positive phages and plasmid colonies were believed to be derived from cDNA synthesised from highly purified extracellular virion RNA in an endogenous RT reaction (Rethwilm et al. 1987). It took more than 10 years of further research to recognise that the probes, as well as some of the first molecular clones, were derived from virion DNA. Much of the early laborious work on obtaining the first FV clones could have been saved had we known that we could simply clone a DNA virus of around 10 kb in length.

In contrast to the blotting and AZT experiments, quantitative competitive PCR and RT-PCR studies yielded heterogeneous results, some of which suggest that virions may actually contain more RNA than DNA (Roy et al. 2003; Yu et al. 1999). Multiple experiments have been performed on two FV isolates from primates and from felines, which are only distantly related (Roy et al. 2003; Yu et al. 1999). The results showed considerable variability in the relation of virion DNA to RNA in the range of approximately 5:1 to 1:5 (Roy et al. 2003; Yu et al. 1999). Some of these findings appear to contradict the above-mentioned experiments on the characterisation of the FV genome.

FV Pcl proteins are not required for the incorporation of viral RNA into capsids (Baldwin and Linial 1999). RNA-containing particles can be exported from infected cells, provided cognate Env glycoprotein is present (Baldwin and Linial 1998; Fischer et al. 1998; Pietschmann et al. 1999). To resolve conflicting results on the nature of the FV genome, I suggest that two kinds of extracellular particles are produced by infected cell cultures. Those with insufficient amounts of Pol protein contain almost exclusively RNA, probably in pieces, which is therefore detectable by RT-PCR but not by Northern blotting. When an attempt was made to characterise the FV genome in the pre-molecular cloning era, researchers, even then, noted the difficulty of detecting full-length virion

RNA compared with smaller pieces (Hruska and Takemoto 1975; Loh and Matsuura 1981). Arguing along these lines, it may be that the integrity of full-length pre-genomic RNA is actually a requirement for the incorporation of sufficient amounts of Pol. The low-Pol-containing viruses are functionally irrelevant for further rounds of infection. The other type of particles contain linear double-stranded DNA. These particles have incorporated full-length RNA and enough Pol proteins in the virus-producing cell and have, therefore, already reverse transcribed their RNA pre-genome. They are the only functionally relevant virions, which maintain the infectious process. A technically demanding, careful analysis of FV gradient fractions should provide proof of this hypothesis, because, in comparison with DNA-containing viruses, RNA-containing viruses should independently migrate on gradients.

4 Integration

Besides reverse transcription, integration as a provirus is the hallmark of the *Retroviridae*, which distinguishes them from the other group of reverse-transcribing viruses, the *Hepadnaviridae* (Brown 1997; Bushman 2002). Retroviral integration is an ordered process which starts with the removal of usually two nucleotides from each 3'OH end of the linear viral DNA (Brown 1997; Bushman 2002). This cleavage reaction leaves recessed ends which ultimately result in the 5' TG and the 3' CA dinucleotides which all proviruses start and end with, respectively (Brown 1997; Bushman 2002). The cleavage reaction is followed by the joining reaction in which the free 3' OH groups are used to attack the phophodiester backbone of the host cell target DNA in a transesterification reaction (Brown 1997; Bushman 2002). All these reactions are carried out by the viral integrase (IN) enzyme and can readily be analysed by in vitro reactions (Brown 1997; Bushman 2002). Because the insertion of the viral DNA occurs at staggered positions of the target DNA, 4–6 bp of the host cell sequence at the integration site are duplicated (Brown 1997; Bushman 2002). After joining the recessed viral 3' ends to host sequences the overhanging and non-base pairing 5' ends are removed on the filling of host sequences (Brown 1997; Bushman 2002). The latter two reactions are probably carried out by the host DNA repair machinery (Brown 1997; Bushman 2002).

FV replication depends on an active IN enzyme, as shown by two reports in which an infectious molecular clone of PFV was rendered replication-deficient after inactivation of the IN catalytic centre (Enssle et al. 1999; Meiering et al. 2000). Because other functions of the viral replication cycle, in particular the generation of virion DNA and its delivery to new cells, were found to remain unaffected by the introduced mutation (Enssle et al. 1999), which affected only a single amino acid, it is fair to say that spumaretroviruses depend on integration just as orthoretroviruses do. However, FVs diverge from the general scheme of integration outlined above in one, probably only minor, aspect. And the reason for this is not understood at all. All FVs bear a peculiar motif at the junction between the downstream *ppt* and the start of the U3 region of the 3' LTR. This makes it very unlikely that the two distal nucleotides are cleaved at the left end of the linear viral DNA molecule to result in the terminal TG dinucleotide that FV proviruses usually start with (Enssle et al. 1999; Juretzek et al. 2003). On the other hand, the structure of the right, U5 end suggests the contrary: that a cleavage had to occur to result in the terminal CA dinucleotide (Enssle et al. 1999; Juretzek et al. 2003). The analysis of two LTR junction sequences of circular DNA molecules, which are generated as dead-end by-products during infection, and of proviral DNA-host cell genome junction sequences, provided formal proof for this unusual feature of the FV integration process (Enssle et al. 1999; Juretzek et al. 2003; Meiering et al. 2000; Neves et al. 1998). Thus we are left with a highly asymmetrical cleavage reaction of the FV IN, which processes the right, U5 end and leaves the left, U3 end intact (Fig. 3). The consequences of this on the integration process are not clear.

The PFV IN protein has been expressed in bacteria, and the cleavage and joining reactions have been analysed in vitro with oligonucleotides as substrate and target molecules (Pahl and Flügel 1993; Pahl and Flügel 1995). Curiously, it was found that the FV IN protein is able to cleave both ends of the linear viral DNA when two nucleotides, which were complementary to the normally cleaved right-end AT dinucleotide, were artificially added to the left-end oligonucleotide (Pahl and Flügel 1993). The cleavage at this modified U3 end was found to be only approximately half as efficient as the regular U5 end cleavage (Pahl and Flügel 1993). However, wild-type FVs do not appear to make use of this.

The cleavage reaction appears to be required for orthoretroviral integration, because the deletion of the two distal U5 end nucleotides has

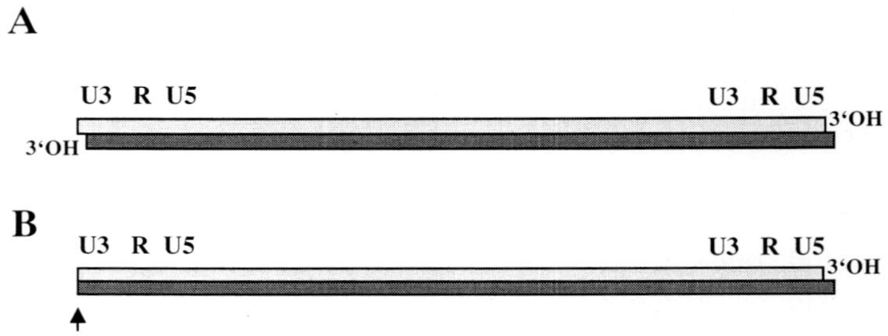

Fig. 3A, B. Retroviral integration. **A** The integrase of orthoretroviruses cleaves usually two nucleotides from each 3'OH end of the linear viral DNA before joining with the host cell genome. **B** FV IN cleaves only the right, U5 end of the DNA, while the left, U3 end (*arrow*) remains unprocessed. The reason for this asymmetrical reaction is not known

deleterious effects on viral replication of MLV (Colicelli and Goff 1985). The two 5' end bases on both sides of the linear viral DNA, which temporarily remain after initial processing of the two 3' end nucleotides, have been implicated in in vitro binding to IN protein (Ellison and Brown 1994). Furthermore, the cleavage reaction creates defined ends on both sides of the provirus even when during reverse transcription untidy ends may have been generated (Bushman 2002). The yeast Ty retrotransposons replicate through an RNA intermediate and integrate into the host cell genome just like retroviral elements (Bushman 2002). Although the principal order of events with respect to reverse transcription and integration is very similar to that of retroviruses, cleavage reactions do not occur on both ends of the linear Ty DNA before integration (Brown 1997). In accordance with the hypothesis of a function of the cleavage reaction in establishing a provirus with defined ends, untidy ends appear to be associated with the generation of Ty replication intermediates (Mules et al. 1998). This, however, does not inhibit Ty integration.

FVs would seem to follow an intermediate pathway in that they cleave on one side and do not cleave on the other. A closer inspection of the process of FV integration and reverse transcription is required to resolve this asymmetry.

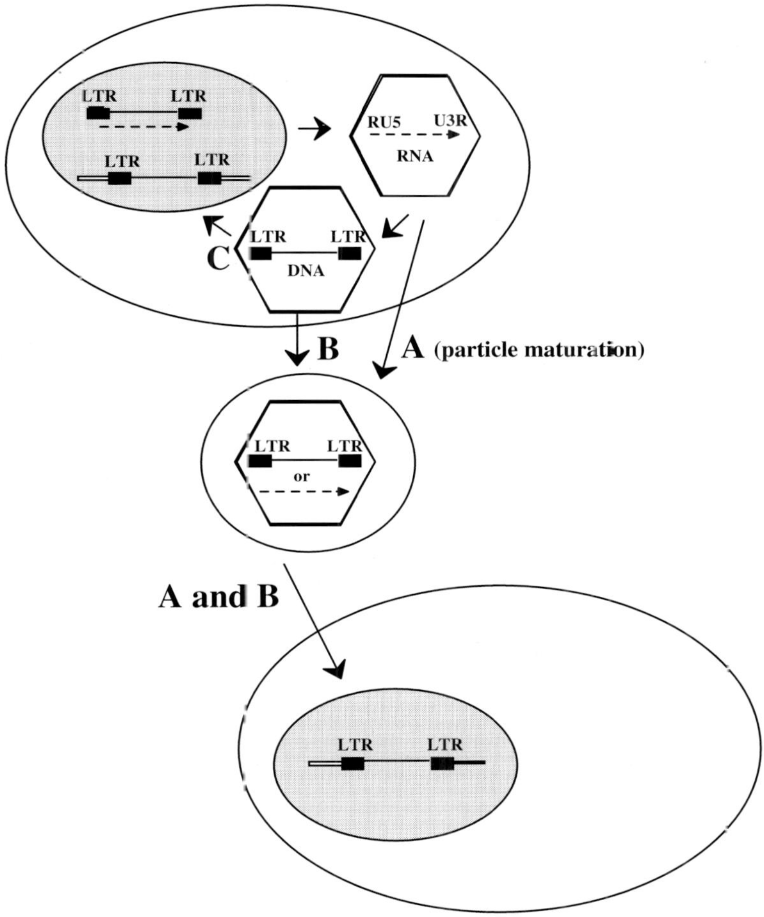

Fig. 4. Principal replication pathway of exogenous orthoretroviruses (pathway *A*) and exogenous (*B*) and endogenous (*C*) pathways of FV. Note that not all orthoretroviruses form intracellularly assembled capsids. This, however, is not of relevance for this figure, because they are essentially RNA viruses and follow an exogenous pathway. (Modified from Heinkelein et al. 2000)

5
Intracellular Retrotransposition

Intracellular retrotransposition (IRT) is defined here as the integration into the host cell genome of cDNA which had been reverse transcribed late in the replication cycle of a retroid element, without having an extracellular phase (Fig. 4, pathway C). IRT is an extremely rare event in orthoretroviruses (Heidmann et al. 1988; Heidmann and Heidmann 1991). Selectable marker genes which were incorporated into suitable orthoretroviral vectors were necessary to detect approximately one event of IRT among 10^6 endogenous viruses per replication cycle (Heidmann and Heidmann 1991). Orthoretroviruses usually require an extracellular phase to activate their reverse transcribing machinery by protease-mediated cleavage of the Gag and Gag-Pol protein precursors, which is the latest event in orthoretroviral replication and leads to particle maturation (Swanstrom and Wills 1997; Vogt 1997a) (Fig. 4, pathway A). These cleavages prepare orthoretroviruses to infect a cell. For the replication of orthoretroviruses it appears not to make sense that an already infected cell acquires multiple proviruses, but to infect further yet uninfected cells. There is indirect evidence for this from HIV, which has developed particular strategies to downregulate the CD4 receptor from infected cells to prevent superinfection from outside (Green and Peterlin 2002). When superinfection occurs under particular cell culture conditions in simpler retroviruses, this may result in the acquisition of multiple copies of unintegrated viral DNA, which can harm the cell (Temin 1988). Thus one or a few intact proviruses per cell seem to be sufficient for orthoretroviruses to maintain an infection (Jung et al. 2002).

The situation is very different in hepadnaviruses. The circular genome of these reverse transcribing viruses, which is densely packed with coding sequences, fails to integrate into the host cell (Ganem and Schneider 2001; Nassal and Schaller 1993). Liver cells, on the other hand, undergo divisions, in particular when chronically infected (Ganem and Schneider 2001). Whereas the proviruses of retroviruses become integrated into the host cell genomes and are inherited by daughter cells, hepadnaviruses are believed to persist by reshuttling their circular DNA into the nucleus to guarantee the infection of daughter cells (Ganem and Schneider 2001; Nassal and Schaller 1993).

The yeast Ty and the *Drosophila* copia elements are well studied examples of class I retrotransposons (Boeke and Stoye 1997; Bushman 2002). These elements have the genetic structure of retroviruses devoid of the *env* gene (Boeke and Stoye 1997; Bushman 2002). Class I retrotransposons jump in the genome by IRT (Boeke et al. 1985; Boeke and Stoye 1997; Bushman 2002; Wilkinson et al. 1994). There is good evidence for the hypothesis that some class I elements can be transferred horizontally to other species, provided an *env*-like function is supplied by a different element (pseudotyping) (Bushman 2002; Jordan et al. 1999).

The usual pathway of FV replication is illustrated by pathway B in Fig. 4. The ability to reverse transcribe the RNA (pre-)genome late in the replication cycle opens up the possibility of shuttling a cDNA copy back into the nucleus and of accumulating further proviruses within a single cell. Thus FVs were thought to combine the strategies of retroviruses (obligatory extracellular phase plus integration) and hepadnaviruses (facultative intracellular phase minus integration) (Heinkelein et al. 2000). In this sense FVs would behave like class I retrotransposons.

Up to 20 proviral copies per genome have been found in persistently FV-infected cells (Meiering et al. 2000). However, the relative contribution of superinfection in the acquisition of these high proviral copy numbers could not be investigated in this study (Meiering et al. 2000). A potential IRT by retroviral genomes has been directly addressed by using *env* gene-deleted vectors of FV, MLV and HIV, which were equipped with the green fluorescent protein marker and able to perform only a single round of transcription, reverse transcription and integration (Heinkelein et al. 2000). Although IRT of MLV and HIV vectors was below the limit of detection by FACS analysis, between 2 and 6% (depending on the cell type) of cells transfected with the FV vector continued to express the marker gene longer than four weeks, indicating stable integration (Heinkelein et al. 2000). The use of AZT as an inhibitor of RT and mutants in the original vector which affected *gag* or *pol* gene expression, or which disabled the activity of RT or IN, followed by sequencing of proviral integration sites demonstrated that a reverse transcription step had, indeed, taken place before the establishment of new proviruses (Heinkelein et al. 2000). The characterisation of individual cell clones by Southern blot revealed that cells harboured from one to more than ten newly acquired proviruses (Heinkelein et al. 2000).

The frequency of IRT in the percentage range seen for FVs is three to four orders of magnitude more than the best results obtained with orthoretroviral constructs (Heidmann et al. 1988; Heidmann and Heidmann 1991; Heinkelein et al. 2000). However, neither the conditions of FV IRT nor its biological function has yet been satisfyingly addressed. As far as the conditions are concerned, it can be assumed that nuclear localisation signals (NLSs) located in the Gag and Pol proteins of the prototypic FV (PFV) are required for the nuclear shuttling of the cDNA complex (Imrich et al. 2000; Schliephake and Rethwilm 1994). There is indirect evidence for this, because the feline FV (FFV) Gag is either not, or very inefficiently, translocated to the nucleus (Bodem et al. 1998; Roy et al. 2003). Moreover, there is no evidence for a FFV IRT (Roy et al. 2003). Thus IRT may be confined to primate FV infections. Furthermore it was shown that a replication-competent PFV with a deletion in the Gag NLS acquired fewer proviral copies in persistently infected cells compared with undeleted virus (Meiering et al. 2000; Yu et al. 1996b). However, this Gag-NLS deletion mutant replicates slightly less well compared with wild-type virus (Yu et al. 1996b). Thus some other defects besides nuclear localisation of Gag cannot be formally excluded. The influence of Env on PFV IRT has been studied only with one mutant, shown to be incapable of cleaving the Env protein into surface and transmembrane subunits (Bansal et al. 2000; Pietschmann et al. 2000). This mutant permits particle export but is unable to infect from outside (Bansal et al. 2000; Heinkelein et al. 2000; Pietschmann et al. 2000). If anything, there was a positive effect of co-expressing this *env* construct together with the PFV vectors on the rescue of cells stably expressing the marker gene (Heinkelein et al. 2000). However, the influence of Env protein on the ability of PFV to retrotranspose within a cell needs to be thoroughly analysed. Furthermore, although some quantification was done in the initial study by Heinkelein et al. (2000), there is a need to quantify exactly how many additional proviral integrations can be generated by primate FV IRT. Only a minority of cells acquired additional proviruses, in some cases more than ten, whereas other cells remained negative. These curious findings await explanation. It may be that the reason for this observation is the experimental design, i.e. those cells which acquired many more additional integrants may have died and thus escaped detection.

Quantification and kinetic studies, together with the analysis of the conditions required for IRT, should facilitate understanding of the biological significance, as well as the potential evolutionary advantage primate FV may have gained from making use of such a mechanism. At the moment one can only speculate on this. Provided further experiments can substantiate FV IRT and can resolve the questions raised above, the following scenario may be considered: the primate FV genome is transcriptionally silent in most infected cells of the body for most of the time (see Falcone et al., this volume). Once the virus starts to replicate it initially performs rounds of replication within infected cells to avoid attack by neutralising antibodies and to accumulate more and more proviruses as matrices for further transcription. Only at a very late stage do primate FVs egress from an infected cell, encounter the hazardous extracellular environment, and leave behind a dying host cell whose genome has been virtually bombarded with proviruses generated through IRT. Obviously, this scenario already addresses the replication strategy of PFV and related primate viruses in vivo, which will by experimentally hard to study. However, some aspects can be clarified by cell culture experiments. Among these is the question of whether there exists any viral or cellular regulation of PFV IRT (see next section).

6
Bet and △Tas Foamy Viruses

Most of the primary transcripts of FVs are efficiently spliced (for details, see chapter by M. Löchelt, this volume). The most efficient splicing event occurs within the region of the accessory genes and involves the *tas* and *bel-2* reading frames (Baunach et al. 1993; Muranyi and Flügel 1991). This leads to the Bet protein (Fig. 1b), which is normally generated from subgenomic transcripts starting at the LTR U3 promoter or, preferentially, the internal *env* gene-located promoter (IP) (Baunach et al. 1993; Löchelt et al. 1994). Bet is probably the protein most highly expressed in FV-infected cells (Baunach et al. 1993), its function, however, is still ill-defined, although there is evidence that in FFV it is required for efficient particle egress (Alke et al. 2001).

In FV-infected cells and in extracellular virions a particular FV genome has been detected, which was derived from a pre-genomic RNA containing the *bet* splice (Saïb et al. 1993). Thus viral genomes were found which were apparently full-length except for a small intron comprising mainly *tas* sequences (Fig. 5). These △Tas viruses are replication-deficient, because they lack essential sequences from the *trans*-activator gene (*tas*). Because cellular conditions which allow the replication of *tas*$^-$ viruses are not known and very unlikely, given the mode of Tas action (see next chapter), the △Tas viruses depend for their spread on either a full-length virus which could act as a helper virus, or any FV deletion variant which is able to express a functional Tas protein. Thus △Tas is analogous to defective interfering viruses known from other systems and may represent a retroviral DI genome (Condit 2001; Huang and Baltimore 1977).

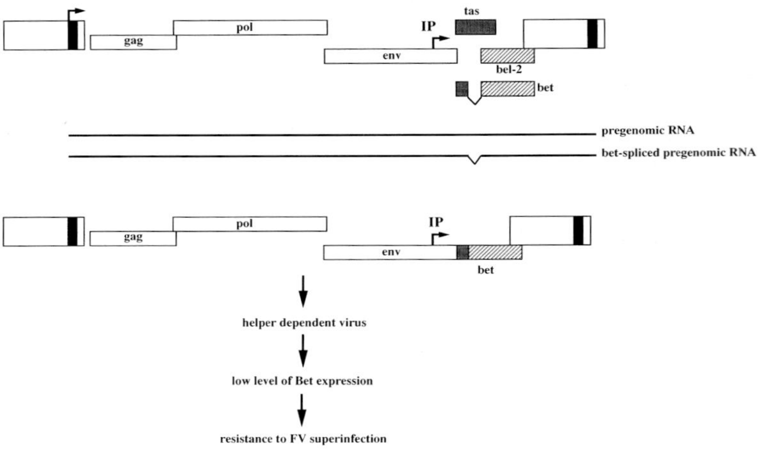

Fig. 5. Generation and structure of the △Tas virus. This virus is derived from an almost full-length RNA pre-genome spliced in the regulatory region. In the △Tas virus the internal promoter (*IP*) is able to drive some expression of the Bet protein which appears to be involved in resistance to superinfection. See text for further details

△Tas viruses have been found to various extents in cell culture, in infected laboratory animals and in natural hosts (Falcone et al. 1999; Meiering wt al. 2000; Saïb et al. 1993; Saïb et al. 1997). However, kinetic analyses of the development of the amount of △Tas viruses in relation to the undeleted form have not yet been performed. Such studies could reveal whether △Tas behaves like other DI viruses and whether an oscillating nature of the amounts of full-length and FV-DI genomes is involved in the establishment of FV persistence (Condit 2001; Huang and Baltimore 1977).

As mentioned above, there is no evidence that △Tas viruses facilitate transcription of structural genes from the LTR U3 promoter. However, △Tas is not transcriptionally silent, because of a low basal IP activity (see next chapter), which may give rise to some Bet protein, although at hardly detectable levels (Saïb et al. 1995). Interestingly, cells harbouring △Tas genomes are refractory to FV superinfection (Saïb et al. 1995). The level of resistance appears to depend on the △Tas copy number integrated into the cell genome, which must be rather high to block infection (Saïb et al. 1995). This suggested a role for Bet in mediating resistance to superinfection (Bock et al. 1998; Saïb et al. 1995). These studies have been taken one step forward by analysis of cell clones which contained and expressed exclusively the viral Bet protein (Bock et al. 1998). These cell clones restricted FV superinfection provided the Bet protein was derived from the superinfecting virus (Bock et al. 1998). Furthermore, the block of productive infection was found to be neither at the level of entry nor at the levels of gene expression (as analysed after transient transfection of plasmids), particle assembly or egress (Bock et al. 1998). These findings made an inhibition after entry and before provirus establishment highly likely (Bock et al. 1998).

Two alternative explanations can be offered for these findings. There is no evidence that Bet influences the transcriptional activity of FV promoters (He et al. 1993; Mergia et al. 1991). However, these analyses have all been made by using transient cotransfection of Bet-expressing and indicator gene plasmids or infectious molecular clones (He et al. 1993; Mergia et al. 1991). It cannot, therefore, be formally excluded that Bet exerts a negative transcriptional effect on the provirus. Because the biological basis of such effects is not obvious, I favour the second explanation. Provided the IRT of primate FVs really takes place at the very efficient level suspected in the preceding paragraph, the virus may have developed inhibitory mechanisms. The Bet protein of primate FVs could inhibit IRT by promoting particle export from the cell rather than allowing backshuttling of the cDNA to the nucleus. Bet is a mainly cytoplasmic protein expressed after provirus establishment and has not been detected as being virion-associated (Baunach et al. 1993; Netzer et al. 1990). Therefore, the attempt to infect already Bet^{+} cells may represent an artificial situation in which Bet probably exerts one of its functions (inhibition of nuclear transport of the capsid or the functionality of the pre-integration complex) at the wrong time of the replication cycle. All the necessary tools to address experimentally this hypothesis are at hand.

It should be mentioned, however, that Bet was recently shown to be secreted from infected cells and taken up by naïve cells (Lecellier et al. 2002). If this turns out to occur at significant levels, it would argue against the above hypothesis. Furthermore, it was shown recently that Bet exhibits a negative effect on the activity of the IP suggesting a role of Bet as a molecular switch to control FV latency and reactivation (Meiering and Linial 2002).

Table 1. Comparison of foamy viruses with orthoretroviruses and hepadnaviruses

	Orthoretroviruses	Foamy viruses	Hepadnaviruses
Genome and expression	ssRNA (diploid)	(Partially) dsDNA (linear)	Partially dsDNA (circular)
	Early reverse transcription (tRNA primer)	Late reverse transcription (tRNA primer)	Late reverse transcription (pol protein primer)
	Provirus integration required LTR-gag-pol-env-LTR (accessory genes)	Provirus integration required LTR-gag-pol-env-LTR plus accessory genes	No integration DR-core-pol-surface-DR (accessory genes)
	Usually one promoter and extensive splicing	Usually two promoters and extensive splicing	More than two promoters and minor splicing events
Capsid and genome packaging	Gag precursor forms VLPs	Gag forms capsids but no VLPs	Core protein forms capsids
	Protease present	Protease present	No protease
	Functionally defined small Gag cleavage products (MA, CA, NC), condensed cores following maturation	Two large Gag proteins infectious virions with immature looking cores	No core protein cleavage
	Cytoplasmic localisation of Gag	Nuclear localisation of Gag, reshuttling of vDNA into the nucleus (primate FVs)	Nuclear localisation of core protein, reshuttling of vDNA into the nucleus
	Efficient pseudotyping with other envelopes	Specific interaction of capsid with Env via LP (ER budding)	Specific interaction of core with S (ER budding)

Table 1. (continue)

	Orthoretroviruses	Foamy viruses	Hepadnaviruses
	gRNA packaged by Gag (Cys-His-boxes)	pgRNA packaged by Gag (Gly-Arg-boxes)	pgRNA packaged by Pol and core
Polymerase	Gag/Pol fusion protein (translational suppression)	Gag-independent expression of Pol (spliced pol mRNA)	Core-independent expression of Pol (internal initiation)
	Pol is incorporated into particle via Gag	Complex interactions of pgRNA, Gag, and Pol, pgRNA required for Pol incorporation	Pol binds to pgRNA through TP
	Integrase present	Integrase present, mechanism of integration differs from orthoretroviruses	No integrase
	Nuclear localisation of integrase	Nuclear localisation of Pol proteins	?

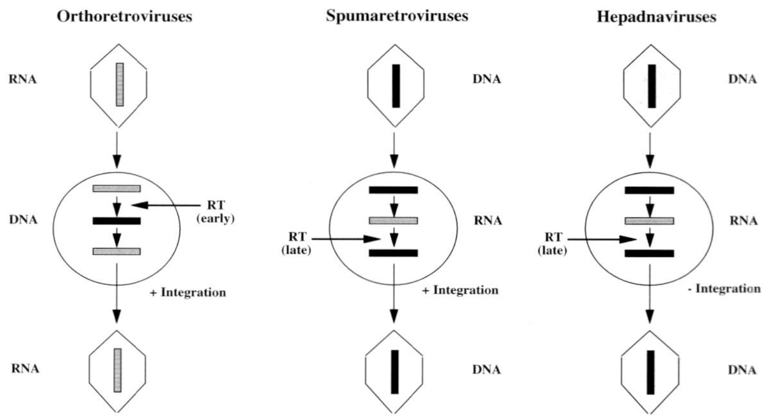

Fig. 6. Schematic view of the most salient features of orthoretrovirus, spumaretrovirus and hepadnavirus replication

7
Comparative Replication Strategies of Retroid Viruses

From the aforementioned discussion, it appears that FVs make use of a replication strategy that shows aspects of the well-established orthoretroviral and hepadnaviral replication pathways. In addition, there are a variety of features unique to this virus group. An overview of the position which certain steps of FV replication hold in relation to orthoretroviruses and hepadnaviruses is shown in Table 1. From a more abstract point of view orthoretroviruses are RNA viruses replicating through a DNA intermediate which has to integrate into the host cell genome, whereas hepadnaviruses are non-integrating DNA viruses replicating through an RNA intermediate. FVs take a middle position, in that they are DNA viruses replicating through an RNA intermediate, which require integration (Fig. 6).

There has been speculation that the reverse-transcribing hepadnaviruses and retroviruses have a common ancestor (Miller and Robinson 1986). Given the FV replication pathway, this is possible. However, it is more likely that FVs evolved independently from the other reverse-transcribing exogenous viruses, because they have acquired so many specialised features which are found neither in orthoretroviruses nor in hepadnaviruses (Table 1). Because viral phylogeny is largely based on

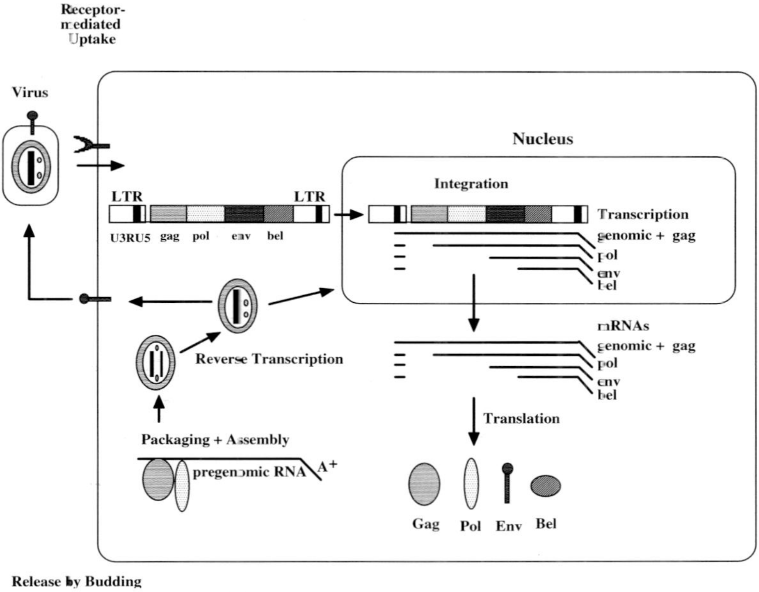

Fig. 7. Overview of the FV replication pathway

the calculation of mutation rates of recent viruses and the comparison of conserved protein motifs or nucleic acid sequences, rather than on the identification of fossils and missing links, any opinion in this area is highly speculative. Some of the FV specialised features have been addressed here, and some are dealt with in the following chapters.

8
Conclusions

FVs appear to pursue a unique pathway of replication with features unprecedented in any other member of the family of *Retroviridae* (Fig. 7). An understanding of some aspects of their replication strategy is now emerging, but the majority still remain to be investigated. More basic research into the molecular biology of FVs is required for the potential use of FV vectors in gene therapy protocols, in addition to allowing us to uncover how flexible viruses and, in particular retroid viruses, maintain and expand their genetic material.

Acknowledgements. I am indebted to Myra McClure for critical review of the manuscript. The research in my lab has been supported over the last years by the Bayerische Forschungsstiftung, Bundesministerium für Bildung und Forschung, Deutsche Forschungsgemeinschaft, European Community and Sächsisches Staatsministerium für Wissenschaft und Kunst.

References

Alke A, Schwantes A, Kido K, Flötenmeyer M, Flügel RM, Löchelt M (2001) The bet gene of feline foamy virus is required for virus replication. Virology 287:310–320

Baldwin DN, Linial ML (1998) The roles of pol and env in the assembly pathway of human foamy virus. J Virol 72:3658–3665

Baldwin DN, Linial ML (1999) Proteolytic activity, the carboxy terminus of gag, and the primer binding site are not required for pol incorporation into foamy virus particles. J Virol 73:6387–6393

Bansal A, Shaw KL, Edwards BH, Goepfert PA, Mulligan MJ (2000) Characterization of the R527T point mutant of a putative cleavage site in human foamy virus env. J Virol 74:2949–2954

Baunach G, Maurer B, Hahn H, Kranz M, Rethwilm A (1993) Functional analysis of human foamy virus accessory reading frames. J Virol 67:5411–5418

Bock M, Heinkelein M, Lindemann D, Rethwilm A (1998) Cells expressing the human foamy virus (HFV) accessory bet protein are resistant to productive HFV superinfection. Virology 250:194–204

Bodem J, Zemba M, Flügel RM (1998) Nuclear localization of the functional bel 1 transactivator but not of the gag proteins of the feline foamy virus. Virology 251:22–27

Boeke JD, Garfinkel DJ, Styles CA, Fink GR (1985) Ty elements transpose through an RNA intermediate. Cell 40:491–500

Boeke JD, Stoye JP (1997) Retrotransposons, endogenous retroviruses, and the evolution of retroelements. In: Coffin JM, Hughes SH, Varmus HE (eds) Retroviruses. Cold Spring Harbor Laboratory Press, Cold Spring Harbor, New York

Brown PO (1997) Integration. In: Coffin JM, Hughes SH, Varmus HE (eds) Retroviruses. Cold Spring Harbor Laboratory Press, Cold Spring Harbor, New York

Bushman F (2002). Lateral DNA transfer. Cold Spring Harbor Laboratory Press, Cold Spring Harbor, New York

Cain D, Erlwein O, Grigg A, Russell RA, McClure MO (2001) Palindromic sequence plays a critical role in human foamy virus dimerization. J Virol 75:3731–3739

Charneau P, Clavel F (1991) A single stranded gap in human immunodeficiency virus unintegrated linear DNA defined by a central copy of the polypurine tract. J Virol 65:2415–2421

Charneau P, Alizon M, Clavel F (1992) A second origin of DNA plus-strand synthesis is required for optimal human immunodeficiency virus replication. J Virol 66:2814–2820

Charneau P, Mirambeau G, Roux P, Paulous S, Buc H, Clavel F (1994) HIV-1 reverse transcription. A termination step at the center of the genome. J Mol Biol 241:651–662

Colicelli J, Goff SP (1985) Mutants and pseudorevertants of Moloney murine leukemia virus with alterations at the integrations site. Cell 42:573–580

Condit RC (2001) Principles of virology. In: Knipe DM, Howley PM (eds) Fields Virology. Lippincott Williams & Wilkins, Philadelphia, PA.

Ellison V, Brown PO (1994) A stable complex between integrase and viral DNA ends mediates human immunodeficiency virus integration in vitro. Proc Natl Acad Sci (USA) 91:7316–7320

Enssle J, Fischer N, Moebes A, Mauer B, Smola U, Rethwilm A (1997) Carboxy-terminal cleavage of the human foamy virus gag precursor molecule is an essential step in the viral life cycle. J Virol 71:7312–7317

Enssle J, Moebes A, Heinkelein M, Panhuysen M, Mauer B, Schweizer M, Neumann-Haefelin D, Rethwilm A (1999) An active human foamy virus integrase is required for viral replication. J Gen Virol 80:1445–1452

Erlwein O, Cain D, Fischer N, Rethwilm A, McClure MO (1997) Identification of sites that act together to direct dimerization of human foamy virus RNA in vitro. Virology 229:251–258

Fischer N, Heinkelein M, Lindemann D, Enssle J, Baum C, Werder E, Zentgraf H, Müller JG, Rethwilm A (1998) Foamy virus particle formation. J Virol 72:1610–1615

Falcone V, Leupold J, Clotten J, Urbanyi E, Herchenröder O, Spatz W, Volk B, Böhm J, Toniolo A, Neumann-Haefelin D, Schweizer M (1999) Sites of simian foamy virus persistence in naturally infected African green monkeys: latent provirus is ubiquitous whereas viral replication is restricted to the oral mucosa. Virology 257:7–14

Ganem D, Schneider RJ (2001) Hepadnaviridae: the viruses and their replication. In: Knipe DM, Howley PM (eds) Fields Virology. Lippincott Williams & Wilkins, Philadelphia, PA.

Green WC, Peterlin BM (2002) Charting HIV's remarkable voyage through the cell: basic science as a passport to future therapy. Nature Med 8:673–680

Guo J, Wu T, Kane BF, Johnson DG, Henderson LE, Gorelick RJ, Levin JG (2002) Subtle alterations of the native zinc finger structures have dramatic effects on the nucleic avid chaperone activity of human immunodeficiency virus type 1 nucleocapsid protein. J Virol 76:4370–4378

He F, Sun JD, Garrett ED, Cullen BR (1993) Functional organization of the bel-1 trans activator of human foamy virus. J Virol 67:1896–190.

Heidmann O, Heidmann T (1991) Retrotransposition of a mouse IAP sequence tagged with an indicator gene. Cell 64:159–170

Heidmann T, Heidmann O, Nicolas J-F (1988) An indicator gene to demonstrate intracellular transposition of defective retroviruses. Proc Natl Acad Sci (USA) 85:2219–2223

Heinkelein M, Pietschmann T, Jármy G, Dressler M, Imrich H, Thurow J, Lindemann D, Bock M, Moebes A, Roy J, Herchenröder O, Rethwilm A (2000) Efficient intracellular retrotransposition of an exogenous primate retrovirus genome. EMBO J 19:3436–3445

Hruska JF, Takemoto KK (1975) Biochemical properties of a hamster syncytium-forming ("foamy") virus. J Nat Cancer Inst 54:601–605
Huang AS, Baltimore D (1977) Defective interfering animal viruses. In: Fraenkel-Conrat H, Wagner RR (eds) Comprehensive Virology. Plenum Press, New York.
Imrich H, Heinkelein M, Herchenröder O, Rethwilm A (2000) Primate foamy virus pol proteins are imported into the nucleus. J Gen Virol 81:2941–2947
Jordan IK, Matyunina LV, McDonald JF (1999) Evidence for the recent horizontal transfer of long terminal repeat retrotransposons. Proc Natl Acad Sci USA 96:12621–12625
Jung A, Maier R, Vartanian J-P, Bocharov G, Jungs V, Fischer U, Meese E, Wain-Hobson S, Meyerhans A (2002) Multiply infected spleen cells in HIV patients. Nature 418:144
Juretzek T, Holm T, Gärtner K, Kanzler S, Herchenröder O, Rammling M, Heinkelein M, Rethwilm A (2003) Foamy virus integration. (manuscript in preparation)
Kupiec JJ, Tobaly-Tapiero J, Canivet M, Santillana-Hayat M, Flügel RM, Périès J, Emanoil-Ravier R (1988) Evidence for a gapped linear duplex DNA intermediate in the replicative cycle of human and simian spumaviruses. Nucleic Acids Res 16:9557–9565
Lecellier C-H, Vermeulen W, Bachelerie F, Giron M-L, Saïb A (2002) Intra- and intercellular trafficking of the foamy virus auxilliary bet protein. J Virol 76:3388–3394
Linial ML (1999) Foamy viruses are unconventional retroviruses. J Virol 73:1747–1755
Löchelt M, Flügel RM, Aboud M (1994) The human foamy virus internal promoter directs the expression of the functional bel 1 transactivator and bet protein early after infection. J Virol 68:638–645
Loh PC, Matsuura FS (1981) The RNA of human syncytial-forming (foamy) virus. Arch Virol 68:53–58
Lori F, Veronese FDM, Vico ALD, Lusso p, M. S. Reitz J, Gallo RC (1992) Viral DNA carried by human immunodeficiency virus type 1 virions. J Virol 66:5067–5074
Maurer B, Bannert H, Darai G, Flügel RM (1988) Analysis of the primary structure of the long terminal repeat and the gag and pol genes of the human spumaretrovirus. J Virol 62:1590–1597
Meiering CD, Comstock KE, Linial ML (2000) Multiple integrations of human foamy virus in persistently infected human erythroleukemia cells. J Virol 74:1718–1726
Meiering CD, Linial ML (2002) Reactivation of a complex retrovirus is controlled by a molecular switch and is inhibited by a viral protein. Proc Natl Acad Sci USA 99:15130–15135
Mergia A, Luciw PA (1991) Replication and regulation of primate foamy viruses. Virology 184:475–482
Mergia A, Shaw KES, Pratt-Lowe E, Barry PA, Luciw PA (1991) Identification of the simian foamy virus transcriptional transactivator gene (taf). J Virol 65:2903–2909
Miller RH, Robinson WS (1986) Common evolutionary origin of hepatitis B virus and retrovirus. Proc Natl Acad Sci USA 83:2531–2535
Moebes A, Enssle J, Bieniasz PD, Heinkelein M, Lindemann D, Bock M, McClure MO, Rethwilm A (1997) Human foamy virus reverse transcription that occurs late in the viral replication cycle. J Virol 71:7305–7311

Mules EH, Uzun O, Gabriel A (1998) In vivo Ty1 reverse transcription can generate replication intermediates with untidy ends. J Virol 72:6490–6503

Muranyi W, and Flügel RM (1991) Analysis of splicing patterns of human spumaretrovirus by polymerase chain reaction reveals complex RNA structures. J Virol 65:727–735

Nassal M, Schaller H (1993) Hepatitis B virus replication. Trends Microbiol 1:221–228

Netzer KO, Rethwilm A, Maurer B, ter Meulen V (1990) Identification of the major immunogenic structural proteins of human foamy virus. J Gen Virol 71:1237–1241

Neves M, Périès J, Saïb A (1998) Study of human foamy virus proviral integration in chronically infected murine cells. Res Virol 149:393–401

Pahl A, Flügel RM (1993) Endonucleolytic cleavages and DNA-joining activities of the integration protein of human foamy virus. J Virol 67:5426–5434

Pahl A, Flügel RM (1995) Characterization of the human spuma retrovirus integrase by site-directed mutagenesis, by complementation analysis, and by swapping the zinc finger domain of HIV-1. J Biol Chem 270:2957–2966

Pfrepper K-I, Löchelt M, Rackwitz H-R, Schnölzer M, Heid H, Flügel RM (1999) Molecular characterization of proteolytic processing of the gag proteins of human spumavirus. J Virol 73:7907–7911

Pietschmann T, Heinkelein M, Heldmann MA, Zentgraf H, Rethwilm A, Lindemann D (1999) Foamy virus capsids require the cognate envelope protein for particle export. J Virol 73:2613–2621

Pietschmann T, Zentgraf H, Rethwilm A, Lindemann D (2000) An evolutionarily conserved positively charged amino acid in the putative membrane-spanning domain of the foamy virus envelope protein controls fusion activity. J Virol 74:4474–4482

Rethwilm A, Darai G, Rösen A, Flügel RM (1987) Molecular cloning of the genome of human spumaretrovirus. Gene 59:19–28

Roy J, Rudolph W, Juretzek T, Gärtner K, Bock M, Herchenröder O, Lindemann D, Heinkelein M, Rethwilm A (2003) Feline foamy virus genome and replication strategy. (submitted)

Saïb A, Périès J, de Thé H (1993) A defective human foamy provirus generated by pregenome splicing. EMBO J 12:4439–4444

Saïb A, Koken MHM, van der Spek P, Périès J, de Thé H (1995) Involvement of a spliced and defective human foamy virus in the establishment of chronic foamy virus infection. J Virol 69:5261–5268

Saïb A, Neves M, Giron M-L, Guillemin M-L, Valla J, Périès J, Canivet M (1997) Long term persistent infection of domestic rabbits by the human foamy virus. Virology 228:263–268

Schliephake AW, Rethwilm A (1994) Nuclear localization of foamy virus gag precursor protein. J Virol 68:4946–4954

Schweizer M, Fleps U, Jäckle A, Renne R, Turek R, Neumann-Haefelin D (1993) Simian foamy virus type 3 (SFV-3) in latently infected Vero cells: reactivation by demethylation of proviral DNA. Virology 192:663–666

Swanstrom R, Wills JW (1997) Synthesis, assembly, and processing of viral proteins. In: Coffin JM, Hughes SH, Varmus HE (eds) Retroviruses. Cold Spring Harbor Laboratory Press, Cold Spring Harbor, New York

Telesnitsky A, Goff SP (1997) Reverse transcriptase and the generation of retroviral DNA. In: Coffin JM, Hughes SH, Varmus HE (eds) Retroviruses. Cold Spring Harbor Laboratory Press, Cold Spring Harbor, New York

Temin HM (1988) Mechanism of cell killing/cytopathic effects by nonhuman retroviruses. Rev Infect Dis 10:399–405

Tobaly-Tapiero J, Kupiec JJ, Santillana-Hayat M, Canicet M, Périès J, Emanoil-Ravier R (1991) Further characterization of the gapped DNA intermediates of human spumavirus: evidence for a dual initiation of plus-strand DNA synthesis. J Gen Virol 72:605–608

Trono D (1992) Partial reverse transcripts in virions from human immunodeficiency and murine leukemia virus. J Virol 66:4893–4900

Vogt PK (1997a) Historical introduction to the general properties of retroviruses. In: Coffin JM, Hughes SH and Varmus HE (eds) Retroviruses. Cold Spring Harbor Laboratory Press, Cold Spring Harbor, New York

Vogt VM (1997b) Retroviral virions and genomes. In: Coffin JM, Hughes SH, Varmus HE (eds) Retroviruses. Cold Spring Harbor Laboratory Press, Cold Spring Harbor, New York

Whitwam T, Peretz M, Poeschla E (2001) Identification of a central DNA flap in feline immunodeficiency virus. J Virol 75:9407–9414

Wilkinson DA, Mager DL, Leong J-A (1994) Endogenous human retroviruses. In: Levy JA (ed) The Retroviridae. Plenum Press, New York

Yu SF, Baldwin DN, Gwynn SR, Yendapalli S, Linial ML (1996a) Human foamy virus replication: a pathway distinct from that of retroviruses and hepadnaviruses. Science 271:1579–1582

Yu SF, Edelmann K, Strong RK, Moebes A, Rethwilm A, Linial ML (1996b) The carboxy-terminus of the human foamy virus gag protein contains separable nucleic acid binding and nuclear transport domains. J Virol 70:8255–8262

Yu SF, Sullivan MD, Linial ML (1999) Evidence that the human foamy virus genome is DNA. J Virol 73:1565–1572

Chapter 2

Foamy Virus Transactivation and Gene Expression

M. Löchelt

Abstract An overview of the pattern and mechanisms of spuma or foamy virus (FV) gene expression is presented. FVs are complex retroviruses with respect to their genetic outfit and the elements used to control and regulate expression of the viral genome. The increased insight into

M. Löchelt (✉)
Abteilung Retrovirale Genexpression,
Angewandte Tumorvirologie, Deutsches Krebsforschungszentrum,
Im Neuenheimer Feld 242, 69009 Heidelberg, Germany
E-mail: m.loechelt@dkfz-heidelberg.de

transcriptional and posttranscriptional mechanisms has revealed that the FVs are distinct, unconventional retroviruses clearly apart from the orthoretroviruses. Although less characterized than the orthoretroviruses, FVs have several unique features that are important for construction and assembly of FV-based vectors for targeted gene delivery and vaccination purposes. Some of these distinguishing features are directly related to the FV-specific mechanisms of gene expression and include (1) the presence of an internal, functional active second transcription unit for expression of the nonstructural genes, (2) the utilization of a subgenomic, spliced transcript for Pol protein expression, and (3) distinct but not yet understood mechanisms for the nuclear exit of defined transcripts and thus an additional level of posttranscriptional control of gene expression. Finally, the interactions of the viral transactivator not only with both viral promoters but also with regulatory elements controlling the expression of defined cellular genes are an important issue with respect to vector development and the apparent apathogenicity of FVs in their natural hosts.

1 Introduction

Spuma or foamy viruses (FVs) are complex retroviruses (Cullen 1991; Löchelt and Flügel 1995; Rethwilm 1995; Linial 1999). This characterization of FVs is, at least in part, based on their mode of viral gene expression. This review focuses primarily on recent findings concerning the complex and "unconventional" nature of FV gene expression and addresses some of the presently open questions. Aspects not covered by this article have been addressed and discussed in depth in a number of previous reviews (Rethwilm 1995; Linial 1999; 2000; Lecellier and Saïb 2000; Meiering and Linial 2001).

FVs share a complex genomic organization. Besides the classic retroviral *gag*, *pol*, and *env* genes, additional genes located between *env* and the 3'-long terminal repeat (LTR), collectively designated *bel* genes, have been identified (Flügel et al. 1987; Fig. 1). Subsequently, the *bel 1/tas* gene of the human FV isolate and other FVs was shown to encode a potent transcriptional transactivator of viral gene expression, the Bel 1/Tas protein (Mergia et al. 1990; Rethwilm et al. 1991; Keller et al. 1991; Winkler et al. 1997). By definition (Cullen 1991), the *bel 1/tas* gene is a regula-

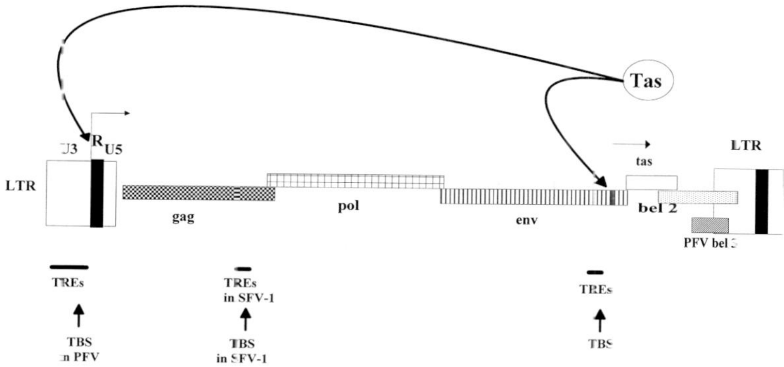

Fig. 1. Schematic overview of the genomic organization of FVs. Genes are depicted as *boxes*, and the LTRs are subdivided into the U3, R, and U5 regions as shown for the 5' LTR. The *bel 3* reading frame is unique for PFV as indicated. The LTR and internal promoters are marked by *rectangular arrows* pointing into the direction of transcription. The localization of Tas response elements (TREs, horizontal bars) and Tas binding sites (TBS, *vertical arrows*) are given; for further details see Lecellier and Saïb (2000). The Tas transcriptional transactivator is schematically shown above the genome with *arrows* pointing to the internal promoter and the LTR promoter to indicate their transactivation by Tas

tory gene because it is absolutely required for viral replication in cultured cells (Löchelt et al. 1991; Baunach et al. 1993; Yu and Linial 1993).

For consistency, the name *tas* for transactivator of spuma virus is used throughout this review, although the acronyms *bel 1* or *taf* (transactivator of foamy virus) are also used for the viral transactivator (Rethwilm 1995; Zou and Luciw 1996). Also for the purpose of consistency in this book, the so-called human FV isolate is designated PFV although the author strongly prefers the original name, human foamy virus (HFV), and the official name will be changed to SFVcpz(hu).

The unique nature of FV gene expression in addition to their complex genomic organization is reflected by the fact that a functionally active internal or intragenic promoter (IP) is located upstream of the regulatory and accessory *bel* genes in the 3'-end of *env* (Löchelt et al. 1993b; Mergia 1994; Campbell et al. 1994; Bodem et al. 1998a; Fig. 1). The IP is efficiently utilized for viral gene expression and required for viral infectivity (Löchelt et al. 1994, 1995). In addition, a *pol*-specific, subgenomic mRNA is made by splicing to direct expression of FV Pol proteins, a mechanism unique to FVs among the retroviruses (Yu et al. 1996a;

Enssle et al. 1996; Bodem et al. 1996). This unusual mechanism of FV Pol protein expression is expected to have profound implications for FV assembly and morphogenesis because a Gag-Pol fusion protein is not formed (Löchelt and Flügel 1996). These features set FV gene expression apart from that employed by other known retroviruses (Coffin 1996).

2
Transcriptional Transactivation of Foamy Virus Gene Expression

The features of the essential FV Tas transcriptional transactivator, the current understanding of its function, and the responsive promoters are described below.

2.1
Description of the Bel 1/Tas Transcriptional Transactivator

None of the FV *bel* genes including *tas* shows strong homology to known cellular genes, nor do they contain sequence motifs that would assign a known enzymatic function to any of them. In the course of genetic experiments primarily on PFV (and SFV-1) Tas using reporter gene assays, site-directed mutagenesis, domain swapping, and biochemical and cell biology techniques, the following characteristics of the FV transactivators have been deduced.

The *tas* gene is located at the 5' end of the *bel* region and overlaps the 3' part of *env* (Fig. 1). The Tas proteins of primate FVs cluster at sizes of about 300 amino acids (aa), whereas bovine (BFV) and equine (EFV) FV Tas consist of 249 aa. The feline FV (FFV) Tas consists of merely 209 aa. There is only a limited degree of homology between the different Tas proteins (Rethwilm 1995; Winkler et al. 1997). This may be because the FV transactivators activate not only the cognate viral promoters but also promoters of defined cellular genes (Wagner et al. 2000). The host species-specific coevolution of the different Tas proteins with defined sets of cellular promoters may have been the driving force for diversification of the FV transactivator proteins.

Among the primate FVs there are some sequence homologies between Tas proteins, but only a few residues are conserved within the whole group of viruses. To identify these conserved residues, the primate and nonprimate (FFV, EFV, BFV) FV Tas sequences were compared in independent alignments. Residues conserved in both alignments were subse-

Tas Domain Structure and Functional Regions

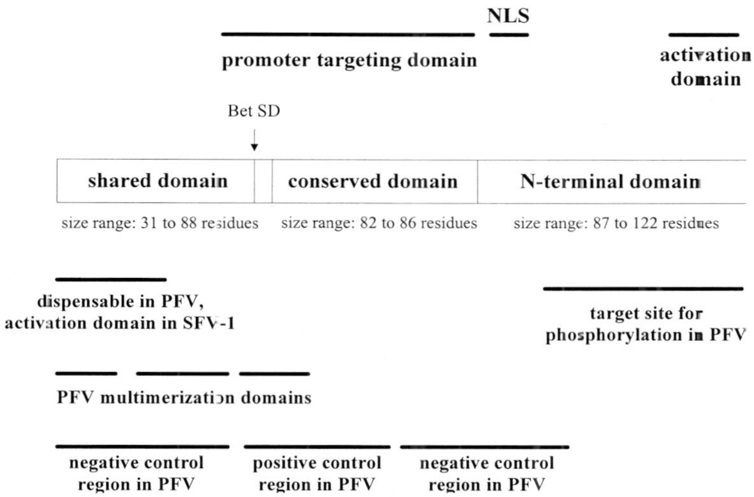

Fig. 2. Schematic summary of FV Tas domain structure and functional regions. According to the sequence alignment presented, the N-terminal shared, the central conserved, and the C-terminal domains of Tas are shown as *boxes*. The range of their sizes is given below the boxes, whereas the position of the Bet SD of PFV, FFV and BFV is given above by a *vertical arrow*. Functional domains of Tas repeatedly and independently identified and described are shown above Tas; those described only once are marked below by *horizontal bars*; in some cases the FV used to define this domain is shown. For further details, see text

quently identified: in the center of the Tas transactivators starting almost directly in that region that is unique for Tas and not contained within Bet (see below), some distinct residues are conserved even among distantly related FVs. This region of 82 to 86 residues in BFV, FFV, and PFV Tas has the following consensus sequence:

$$L-X_{(13)}-I-X_{(18-22)}-Y-X-C-X_{(2)}-C-X_{(5)}-D/E-X_{(10-11)}$$

$$-I/L-X_{(8)}-G-X_{(8-10)}-R/K-H-X_{(3-5)}-C.$$

These residues are primarily localized to a region required for protein-protein interaction and promoter target site specificity (He et al.

1993; Venkatesh et al. 1993; Mergia et al. 1993; Lee et al. 1994b, Chang et al. 1995; Fig. 2). This conserved region is flanked at the N terminus by a domain that is also present in Bet. This shared region is 88 aa long for PFV Tas (Muranyi and Flügel 1991) and only 31 and 33 aa long for Tas proteins of FFV and BFV, respectively (Renshaw and Casey 1994b; Bodem et al. 1998a). The size of the shared N-terminal region showing almost no sequence conservation is in part responsible for the different sizes of the FV Tas proteins.

In PFV, three distinct domains in the N-terminal shared region together with adjacent residues contribute to and control multimerization of Tas (Chang et al. 1995). Although part of this domain is almost dispensable for transactivation in PFV, this domain in SFV-1 contributes to transactivation by providing an additional activation function (He et al. 1993; Venkatesh et al. 1993; Mergia et al. 1993).

The PFV and FFV Tas proteins are located in the nucleus (Keller et al. 1991; Bodem et al. 1998b), as anticipated from their function as transcriptional transactivators (Rethwilm et al. 1991). A complex bipartite nuclear localization signal (NLS) has been experimentally identified in PFV and predicted for the other FVs (He et al. 1993; Venkatesh et al. 1993; Chang et al. 1995). The position of the NLS within Tas appears to be flexible, and its primary sequence is only slightly conserved (Winkler et al. 1997; Tobaly-Tapiero et al. 2000).

Flanked by the N-terminal shared domain and the NLS, a central sequence of about 100 aa appears to target Tas to its DNA binding sites in the Tas-responsive promoters (see below). This protein sequence does not contain any of the DNA-binding motifs known from other proteins that specifically interact with DNA. This functionally important part of the FV transactivators has not been characterized in detail. It is not yet clear whether the DNA binding domain is functionally separated from the NLS, with its enrichment of positively charged residues that could be expected to contribute to DNA binding.

Sequences comprising the central part of Tas have been shown to transactivate the PFV LTR promoter to a low degree; for full transactivation, sequences from the C terminus of Tas are also required (Venkatesh et al. 1993; He et al. 1993).

The C terminus of Tas is characterized by the presence of hydrophobic and acidic residues. A short and conserved N-terminal domain of about 30 residues is absolutely required for transactivation and was shown to function as a modular transactivation domain in a heterologous context, even in yeast (Garrett et al. 1993; Mergia et al. 1993; Venkatesh and Chinnadurai 1993, Venkatesh et al. 1993; Blair et al. 1994; Lee et al. 1994b). This led to the conclusion that Tas interacts with highly conserved component(s) of the cellular transcriptional machinery and allowed further functional studies in heterologous genetic systems (Garrett et al. 1993). Interestingly, these conclusions are underlined by a genetic screen in yeast where the ADA2 transcriptional adapter was shown to be required for the function of the activation domain (Blair et al. 1994). Additionally, a human ADA2 homolog has been identified as a component of the ADA transcription complex implicated in interactions with the basal transcription machinery and in histone acetylation (Carter et al. 1997). These findings are reminiscent of those reported for the acidic activation domain of the VP16 transactivator of herpes simplex virus (He et al. 1993, Blair et al. 1994).

The PFV Tas protein is a phosphoprotein, and data suggest that C-terminal sequences are primarily involved in phosphorylation (Venkatesh et al. 1993). Although the biological function of Tas phosphorylation and its influence on Tas function have not been studied so far, this modification could regulate biological activity. The positive and negative effects of defined PFV Tas sequences on transactivation (Lee et al. 1994b) remain to be further elucidated in order to establish the presence of autoregulatory circuits.

It was recently shown that the cellular promyelocytic leukemia (PML) protein binds PFV Tas, thus interfering with the transactivation function of Tas. It was shown that the association of Tas with PLM inhibited the functionally important DNA-binding activities of the transactivator (Regad et al. 2001). This interaction requires N-terminal Tas sequences and is responsible for the interferon-induced antiviral state resulting in a reduced viral gene expression and infectivity. In addition, Bet was also reported to interact with PML (Regad et al. 2001). However, PML has been shown by others to not contribute to viral latency (Meiering and Linial 2003).

2.2 Mode of Bel 1/Tas-Mediated Transcriptional Transactivation

The Tas proteins of PFV, SFV-1, and FFV (Campbell et al. 1996; He et al. 1996; Zou and Luciw 1996; Kang et al. 1998; Kang and Cullen 1998; Kido et al. 2002; Bodem and Flügel, unpublished data) directly bind to their

DNA target sequences in the viral promoters in vitro. The direct binding of the Tas transactivator to the response elements on the proviral DNA distinguish FVs from HIV-1 Tat, which binds to the TAR RNA response sequence on the nascent mRNA chain, and is also distinct from HTLV Tax, where DNA binding is indirect. Similar to HTLV-I Tax, Tas acts primarily but not exclusively upstream of the cap site but the mechanisms of targeting may be different in HTLV and FVs (Cullen 1992).

For SFV-1, Tas binding sites (TBS) have been identified directly upstream of the IP and in the *gag* gene (Zou and Luciw 1996; Campbell et al. 1996; Fig. 1). In contrast, the PFV TBSs are both upstream of the internal and the LTR promoters (He et al. 1996; Kang et al. 1998; see below). Comparing the viral TBSs of PFV with each other and those of SFV-1, only a low degree of homology is detectable, and comparisons between both viruses yield almost no sequence conservation (Campbell et al. 1996; Kang et al. 1998; Kang and Cullen 1998). This finding has been further extended to SFV-1 by using a genetic screen in yeast: the SFV-1 *gag* TBS showed a strong homology to a genetically identified high-affinity consensus sequence whereas only the core sequence of the IP TBS was conserved (Kang and Cullen 1998). This genetic screen for SFV-1 appears to contradict biochemical and functional studies on the PFV promoters: in PFV, the TBS in the IP was shown to have a much higher affinity to Tas than the TBS located in the PFV LTR promoter (Kang et al. 1998; Fig. 1). Finally, the TBSs for the cellular p57Kip2 gene are located in the coding sequence of the gene consisting of conserved repeats (Kido et al. 2002; see below).

The mechanism of FV transactivation is not presently known. PFV Tas-mediated transactivation of viral gene expression apparently depends, at least in part, on cellular factors (He et al. 1996). At least some of the cellular cofactors required appear to be ubiquitous because Tas of primate FVs is functionally active even in yeast and many different cell types support productive FV replication (He et al. 1996; Kang and Cullen 1998). As discussed above, the hADA2 transcriptional adaptor protein may be a potential Tas-interacting protein required for transactivation (Carter et al. 1997). The critical involvement of cellular factors is further supported by recent findings that both PFV promoters show a considerable degree of cell type specificity with respect to basal activity and Tas transactivation (Linial 2000, Meiering et al. 2001). In particular, cells that are either fully permissive toward PFV infection or allow only a persistent infection show different regulation of gene expression by the LTR and IP (Meiering et al. 2001, see below). Further studies are required to fully define and characterize Tas-mediated activation of gene expression.

The genetically identified Tas response elements (TRE), which mediate Tas transactivation of homologous and heterologous promoters, and the TBS identified biochemically do not necessarily have to be identical or even overlapping (see below). This may indicate that FV transactivators have alternative pathways to activate the cellular transcription machinery: as DNA-binding proteins and/or as proteins that are specifically recruited to the response elements by different means, for instance, by protein-protein interaction with already DNA bound cellular transcription factors as discussed for HTLV-I or by binding to adjacent response elements in a complex with cellular transcription factors (Cullen 1992).

In summary, the domain structure and function of FV transactivators are complex and far from being fully understood.

3
Overview of the FV Promoters and Phylogenetic Considerations

One of the features that sets FVs apart from other known retroviruses is the fact that they possess two different promoters (Fig. 1). The classic retroviral LTR promoter directs expression of viral gene products and, potentially, also of cellular genes that are located downstream of the integrated provirus. In FVs, the 5' LTR has the capacity to direct the expression of *gag*, *pol*, *env*, and *bel* genes. In addition, a strong second promoter is located in the 3'-end of the *env* gene directly upstream of the regulatory and accessory *bel* genes, the IP of FVs. Because of its location downstream of the structural genes, the IP directs accessory and regulatory *bel* gene expression only. Indications for the presence of an additional promoter inside of *pol* upstream of the reverse transcriptase domain have been described for SFV-1 (Campbell et al. 1996) but have not been confirmed by others.

Because none of the FV *bel* genes shows strong homology to regulatory or accessory genes of other retroviruses and because the IP of FVs is unique among retroviruses, the *bel* gene(s) together with its/their promoter may have been incorporated into the genome of the ancestor of the present-day FV. In line with this, the IP and the *tas* gene are functionally linked: Tas is the transcriptional transactivator of the IP (Löchelt et al. 1993b). This may imply that both genetic elements were incorporated as an intact functional unit to enhance and activate FV gene expression. Subsequently, the LTR promoter was also taken under the positive control of Tas, leading to the present-day situation: both promoters are strongly dependent on Tas, and Tas has even become indispensable for viral gene expression and infectivity (Löchelt et al. 1993b, 1995).

Both promoters have consensus TATA boxes upstream of their transcriptional start sites. The start site of transcription has been mapped by primer extension and other techniques and corresponds in most cases to the sequence 5'-GAGCTC-3' or derivatives thereof, the first G nucleotide corresponding to the cap site. Upstream of the TATA box potential binding sites of well-defined transcription factors are present. In PFV, AP-1 consensus binding sites have only a moderate role in gene expression and no role in phorbol ester activation (Maurer et al. 1991; Meiering and Linial, personal communication). In BFV, several binding sites for cellular transcription factors have been identified in computer searches; however, their functional significance has not been experimentally tested (Renshaw and Casey 1994a).

3.1
Structure of the Foamy Virus 5' LTR Promoter

In permissive and also semipermissive cells, the LTR promoters of those FVs analyzed so far exhibit a very low basal activity in the absence of the Tas transactivator when reporter gene constructs containing the complete LTR, or only the U3 domain of the LTR, are studied (Mergia et al. 1990, 1991; Rethwilm et al. 1991; Keller et al. 1991; Winkler et al. 1997, Tobaly-Tapiero et al. 2000; Meiering et al. 2001). Primarily, the *cis*-acting sequences required to confer Tas responsiveness to a defined promoter as determined in genetic assays, i. e., the Tas response elements, TREs, are located upstream of the TATA box with the exception of a strong TRE/TBS in the SFV-1 *gag* gene (Campbell et al. 1996). This SFV-1 TRE/TBS has been mapped to the 3' end of *gag* upstream of *pol* (Campbell et al. 1996). In addition to this *gag* TRE/TBS, the SFV-1 U3 sequence contains two enhancer-like TREs between 1196 to 880 and 403 to 125 relative to the transcriptional start site (Mergia et al. 1992). In SFV-3, the region between 637 and 180 was shown to contain two TREs (Renne et al. 1993). In SFV-1, SFV-3 and PFV, the R-U5 region and also sequences upstream of the *gag* gene have moderate or strong negative effects on gene expression when assayed in heterologous reporter gene constructs or in the context of truncated proviral systems (Mergia et al. 1992; Erlwein and Rethwilm, 1993; Renne et al. 1993; Yang et al. 1997).

The most complex arrangement of TREs has been reported for the PFV U3 promoter: sequences from 500 to 50 relative to the transcriptional start site are required for efficient Tas transactivation, and, depending on the study, two to five distinct TREs have been mapped in this region (Venkatesh et al. 1991; Erlwein and Rethwilm 1993; Lee et al. 1993; summarized by Lecellier and Saïb 2000). In general, these TREs function in an orientation-independent manner reminiscent of cellular enhancer elements. Negative *cis*-acting elements in the PFV U3 have been also described but are less characterized (Erlwein and Rethwilm 1993; Lee et al. 1993; Yang et al. 1997).

The promoter-proximal PFV TRE located ~60 nt upstream of the cap site overlaps with a PFV DNA sequence at about 73 to 47 that is specifically bound by PFV Tas (He et al. 1996; Kang et al. 1998). Tas binding studies combined with genetic data to identify TREs indicate that an unknown cellular factor(s) actively contributes to Tas-mediated transactivation (He et al. 1996; Kang et al. 1998). Similar studies have been used to identify and characterize the TRE in the SFV-1 *gag* gene. A DNA sequence of 40 nt located in the end of *gag* is specifically bound by recombinant SFV-1 Tas (Campbell et al. 1996). A genetic screen in yeast designed to detect optimal SFV-1 Tas binding sites fully confirmed this sequence in *gag* (see below).

The promoter-proximal TREs appear to be of primary importance, because the apparent deletion of a substantial part of U3 sequences in PFV (Herchenröder et al. 1995; Schmidt et al. 1997) does not interfere with PFV gene expression in cell cultures, analogous to corresponding data from studies of the FFV LTR promoter (Schwantes et al. 2002).

3.2 Structure of the Foamy Virus Internal Promoter

As described for the LTR promoter, the localization of genetically identified TRE sequences in the IP does not necessarily coincide with TBS defined biochemically in binding studies: a TRE defined by deletion mutants actually flanked but did not include the TBS of the PFV IP (Yang et al. 1997; Kang et al. 1998). Similarly, two TREs have been genetically mapped to the IP of SFV-1 but only the promoter-proximal site is actually bound by the transactivator in binding assays (Zou and Luciw

1996). Importantly, the IP TBS of SFV-1 has limited homology to the optimal binding sequence as determined in a yeast system and may thus not represent a high-affinity TBS (Kang and Cullen 1998). Apparently in contrast with these results, the same authors (Kang et al. 1998) showed that the PFV IP TBS has a much higher affinity to Tas than the TBS of the LTR promoter. This discrepancy may be related to the fundamentally different nature of the assay systems used or to the fact that unknown cellular factors primarily mediate Tas binding.

The basal, Tas-independent activity of the IP is generally higher than that of the LTR (Löchelt et al. 1993a; 1994; Meiering et al. 2001). It is further increased by the presence of the LTR promoter in *cis.* Because the basal activity of the IP is enhanced, the IP may be the early promoter for FV gene expression that provides the essential Tas transactivator during replication (Löchelt et al. 1993a; 1994; 1995; Meiering et al. 2001; Meiering and Linial 2002). These data are supported by RNA expression studies that show an early expression of (IP-directed) *bel* mRNAs (Mergia 1994; Löchelt et al. 1995).

Interestingly, a cryptic promoter is located 32 to 35 nt upstream of the authentic IP and corresponding transcripts are detectable even when the IP TATA box is intact. Upon mutagenesis of the IP TATA box, the cryptic promoter directs considerable mRNA synthesis (Löchelt et al. 1995). The importance of the IP for virus infectivity through expressing sufficient amounts of the Tas transactivator has been demonstrated by site-directed mutagenesis of the TATA box (Löchelt et al. 1995).

Recently, binding of nuclear factor I (NFI) to the IP of PFV was shown to repress Tas-mediated transactivation and NFI over-expression strongly decreased PFV gene expression (Kido et al. 2003).

These data demonstrate that the FV LTR and the IP promoters are structurally distinct and thus allow differential gene expression with two independent transcription units.

4
Subgenomic Viral Transcripts

The splicing patterns of FVs are complex, and the utilization of a spliced transcript for FV Pol expression adds another layer of complexity to FV gene expression (Yu et al. 1996a; Bodem et al. 1996; Jordan et al. 1996; Holzschu et al. 1998; Figs. 3 and 4).

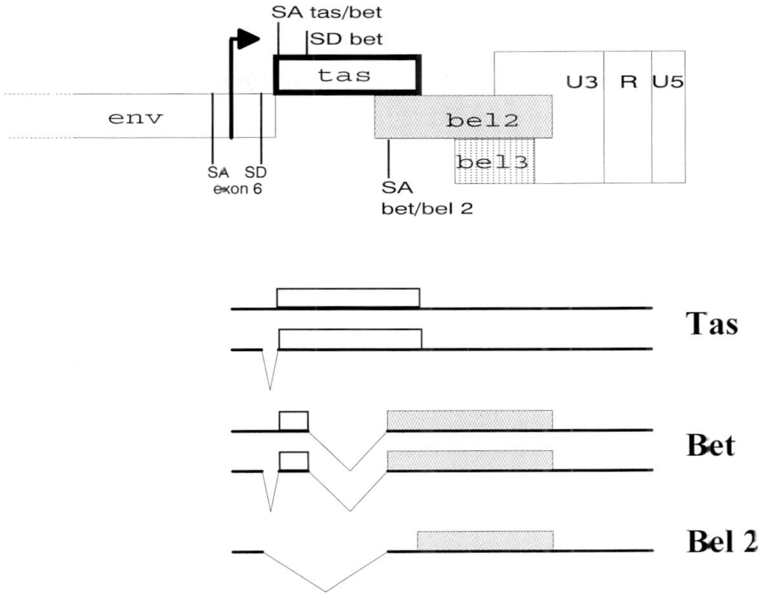

Fig. 3. Schematic presentation of FV IP-directed *bel* transcripts. Above, an enlarged overview of the 3' end of the PFV genome with the *env* and *bel* genes, the 3' LTR, and the IP (*rectangular arrow*) are shown. Relevant splice donor (*SD*) and splice acceptor (*SA*) sites are shown, exon 6 refers to a noncoding exon of LTR-derived PFV *bel* transcripts. Below, the IP-directed *bel* transcripts are shown as *solid lines*, splice events are marked by *thin lines*, and coding sequences are marked by *boxes* corresponding to the genes (Löchelt et al. 1994). The coding capacity of the different transcripts is at the right-hand margin

The presence of two transcription units for FV gene expression distinguishes two basically different transcript classes: those derived from the LTR promoter and, in addition, IP-derived mRNAs exclusively encoding Bel proteins (Löchelt et al. 1994).

4.1
Spliced Transcripts from the Internal Promoter

IP-derived mRNAs are efficiently spliced, and only a minor fraction remains unspliced (Fig. 3). The unspliced IP mRNAs encode the Tas transactivator. However, Tas is primarily encoded by spliced IP transcripts (Löchelt et al. 1994). An intron of ~120 nt that exactly codes for the Env membrane spanning domain (MSD) is spliced out, fusing the ~120-nt-long noncoding leader exon of the IP to the *tas* sequence (Muranyi and Flügel 1991; Löchelt et al. 1993b, Bodem et al. 1998a). Because of its location and strong sequence conservation among different FVs, the removal of this intron appears to have multiple functions: (1) it is used to form spliced *tas/bet* mRNAs; (2) it serves to remove the Env MSD from alternative Env forms (Lindemann and Rethwilm 1998); and (3) this splicing event may serve as a general mechanism to activate and enhance FV gene expression (see below).

The second splice event consistently found in the analyzed FVs is located downstream in *tas,* joining the 5'-end of *tas* in frame to *bel 2,* which leads to the expression of the Bet fusion protein. This splice event is extremely efficient in FV-infected cells and will be designated *bet*-splice (Muranyi and Flügel 1991; Löchelt et al. 1993b, Bodem et al. 1998a). In general, the *bet* splice acceptor (SA) is located far upstream of the Bel 2 initiation codon. Because Bet completely encompasses Bel 2, the identification of Bel 2 proteins in infected cells is difficult and deletions introduced into the FFV *bel 2* sequence affect primarily Bet and not Bel 2 (Alke et al. 2001). Finally, FV sequences located between the splice donor (SD) terminating the IP leader exon and the SA of *bet* can be spliced out, resulting in a Bel 2-encoding transcript of the IP (Muranyi and Flügel 1991; Bodem et al. 1998a; Fig. 3)

4.2
Spliced *gag*, *pol*, and *env* Transcripts from the 5' LTR Promoter

The 5' LTR promoter directs the expression of a full-length genomic mRNA that can be packaged into FV particles (Baldwin and Linial 1998) or used directly or after splicing as a template for translation (Muranyi and Flügel 1991; Bodem et al. 1998a; Fig. 4). It is not fully understood for any retrovirus (including FVs) whether and how control mechanisms exist for executing the multiple decisions on splicing, direct translation, and/or genome packaging (Butsch and Boris-Lawrie 2002). Whereas transcription is a nuclear process, ribosomal translation and retroviral packaging are considered to be localized to the cytosol. A nuclear targeting of part of PFV Gag had been described and discussed as a mechanism for selecting genomic RNA for packaging (Schliephake and Rethwilm 1994). Such an essential nuclear function of Gag in RNA selection and/or RNA transport is, however, contradicted by the observations that

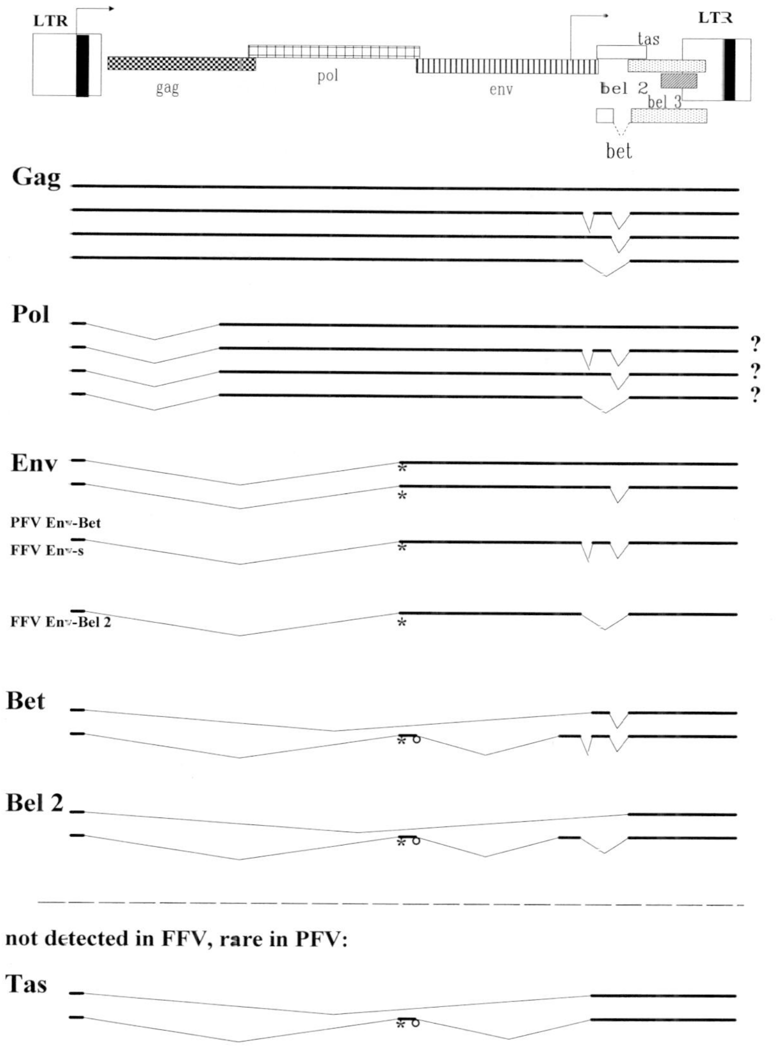

Fig. 4. Schematic presentation of FV LTR-directed transcripts. *Above*, an overview of the complete PFV genome with the *gag, pol, env,* and *bel* genes, the LTR, and IP (marked as *rectangular arrows*) are shown. *Below*, the LTR-directed transcripts are shown as *solid lines* and splice events are marked by *thin lines*. The coding capacity of the different transcripts is at the left-hand margin. The *asterisks* and *circles* mark multiple SA and SD sites located in the center of the genome and used for defined *env* and *bel* transcripts. *Question marks* indicate that the corresponding multiply spliced *pol* mRNAs have not been detected but are postulated to exist (for details, see text)

an PFV mutant that codes for an NLS-deleted Gag is almost fully infectious and viable (Yu et al. 1996b).

As mentioned for IP-derived FV mRNAs, only a small portion of the LTR transcripts remains unspliced (Fig. 4); the majority is either singly or multiply spliced (Muranyi and Flügel 1991; Bodem et al. 1998a). In general, certain SD and SA sequences remain in LTR-derived transcripts and may thus interfere with the nucleo-cytoplasmic export of these partially spliced transcripts (Cullen 2000b).

The major 5' SD of FV LTR transcripts is located between 51 and 57 nt downstream of the 5' LTR cap site and follows closely the splice consensus sequences (Muranyi and Flügel 1991; Bodem et al. 1996; Heinkelein et al. 2000). To express spliced *pol* mRNA, this SD is fused to a single SA in *gag* about 500 (PFV), 300 (FFV), and 240 nt (BFV) upstream of the start codon of *pol* (Yu et al. 1996a; Jordan et al. 1996; Bodem et al. 1996; Holzschu et al. 1998). The *pol* mRNAs of FFV and BFV have been shown to be formed at a high frequency: compared to unspliced *gag*/genomic RNAs, spliced FFV *pol* RNA is approximately half of that of *gag*/genomic RNA as determined by Northern blotting with gene-specific probes (Bodem et al. 1998a). Almost identical amounts of both RNA classes have been detected by quantitative RT-PCR-based techniques in BFV (Holzschu et al. 1998). The comparably high expression of *pol* mRNAs (and proteins) may be a mechanism of FVs to counteract a low efficiency of Pol packaging compared to that achieved by Gag-Pol fusion proteins of other retroviruses. This assumes that the relatively high levels of *pol* mRNAs are efficiently and correspondingly translated into Pol proteins. Surprisingly, PFV Gag (and Pol) expression almost completely depends on the presence of the major SD or a heterologous SD site in the untranslated leader in the context of subgenomic expression plasmids (Heinkelein et al. 2002).

Up to four different SA sites either directly upstream or more distantly spaced from the start of the Env Met start codon (asterisks in Fig. 4) are used to generate *env* mRNAs (Muranyi and Flügel 1991; Bodem et al. 1998a). In comparison to *pol* mRNAs, *env* transcripts are significantly less abundant in FFV-infected cells (Bodem et al. 1998a).

A substantial fraction of structural gene transcripts is further spliced in the *bel* region (Saïb et al. 1993; Lindemann and Rethwilm 1998; Giron et al. 1998; Bodem et al. 1998a). It is unclear to what extent FV *gag*-, *pol*-, and *env* mRNAs are spliced at SD/SA sites in the *env-bel* region as described above for IP-derived mRNAs. At present there are no data on the relevance and biological significance of these additional splice events. Provided that FV transcripts do not function as bicistronic messages, these splices do not change the primary coding capacity of *gag* and *pol* mRNAs. In the case of the *gag*/genomic mRNA, however, these splicing events may gain biological significance: when almost complete FV transcripts that are spliced in the *env-bel* region are packaged, these genomes must be considered as functionally defective because they cannot direct the synthesis of infectious progeny because of the deletion of essential parts of Env and/or Tas (Fig. 4; Saïb et al. 1993). The genomes are even defective in viral gene expression in vivo and in vitro because they lack the Tas transactivator. The defective genomes have been detected in PFV- and FFV-infected cells and in SFV-infected African green monkeys and are suspected of playing a role in the establishment or maintenance of viral persistence (Saïb et al. 1997; Bodem et al. 1998a; Falcone et al. 1999). Because of the premature initiation of cDNA synthesis in FVs (Moebes et al. 1997), spliced genomic transcripts are easy to detect, since they are reverse transcribed into DNA and thus are directly observable (Saïb et al. 1993, 1997; Bodem et al. 1998a; Falcone et al. 1999).

Corresponding splicing events in the *env-bel* region have also been described for PFV and FFV *env* transcripts that express distinct, completely different Env-related proteins (Fig. 4). The utilization of the SD located directly downstream of the IP in *env* results in the removal of the Env MSD. In PFV, the truncated Env is fused in-frame to Bet to form an Env-Bet protein that has been shown to be targeted to the secretory pathway (Lindemann and Rethwilm 1998; Giron et al. 1998). The function of Env-Bet is not known and a genetically modified virus unable to express it is fully infectious in cell cultures (Lindemann and Rethwilm 1998). In the infected host, however, Env-Bet may have important functions, for instance, as a soluble decoy for the immune system. In FFV, minor sequence differences in the location of the SA sequences upstream of *tas* lead to the generation of two distinct proteins (Bodem et al. 1998a). The SD in *env* is fused either directly in-frame to the short cytoplasmic domain of Env, resulting in an Env protein that may exit the cell via the secretory pathway, or, alternatively, Env is fused in-frame onto *bel 2* to express an Env-Bel 2 fusion protein (Bodem et al. 1998a; Fig. 4). In FFV, nothing is known about the biological significance of these Env derivatives but their function in the infected host could be analyzed in cats, the appropriate animal model to analyze the function of individual FV genes (Alke et al. 2000). Secreted FFV Env proteins may be even involved in renal failure due to the deposition of immune complexes in the kidney of FFV-infected cats, as indicated by a recent study (Winkler et al. 1999).

The multiply spliced *env* RNAs are not reverse transcribed in infected cells by the FV reverse transcriptase, indicating that sequences exclusively contained in *gag*/genomic transcripts are essential for cDNA synthesis. This may be related to a requirement for the primer binding site and RNA genome packaging for reverse transcription.

It is not known yet whether *pol* transcripts are also spliced in the 3' region of the mRNA, analogous to *env* transcripts. If such multispliced *pol* transcripts do exist, they are not reverse transcribed in FFV (unpublished data). In any case, such mRNAs should not have an altered coding capacity.

FVs apparently do not encode a posttranscriptional regulator that actively exports unspliced or partially spliced mRNAs out of the nucleus into the cytosol for protein biosynthesis like the HIV Rev and the HTLV-I Rex proteins (Yu and Linial 1993; Baunach et al. 1993; Adachi et al. 1995). Thus splicing may be the mechanism FVs use to escape nuclear retention of partially spliced mRNAs or to counteract nuclear retention signals located in the transcripts (Cullen 2000b) as discussed below.

4.3
Spliced *bel* Transcripts from the 5' LTR Promoter

Besides transcripts that code for the structural Gag, Pol, and Env proteins, Bel expression is also directed from the LTR promoter. It is unclear, however, whether this LTR-derived *bel* gene expression is of biological significance: PFV proviruses with a functionally crippled IP display a strongly reduced viral infectivity. This implicates that adequate *bel* gene expression cannot be fully provided by the LTR promoter (Löchelt et al. 1995). In PFV and FFV, *bel*-specific transcripts have been attributed to either the 5' LTR promoter or the IP (Muranyi and Flügel 1991; Löchelt et al. 1994; Bodem et al. 1998a); this distinction has not been included in transcriptional mapping of BFV (Renshaw and Casey 1994b).

In PFV and BFV, LTR-derived transcripts coding for Tas, Bet, and Bel 2 have been identified; however, reverse transcription-PCR was required for their detection (Muranyi and Flügel 1991; Renshaw and Casey 1994b). In FFV a single LTR-derived multiply spliced Bet mRNA was detectable with this highly sensitive technique; other *bel* mRNAs including those for the Tas transactivator were solely derived from the IP (Bodem et al. 1998a). Comparably, IP-derived PFV *tas* RNA was clearly detectable by primer extension whereas corresponding transcripts of similar size from the PFV LTR were undetectable with this technique, even when the IP was functionally inactivated (Löchelt et al. 1993b, 1995). In summary, the FV Tas transactivator is almost, if not completely (FFV), expressed from the IP and not from the 5' LTR in line with genetic data that show that the IP is required to provide the crucial Tas protein (Löchelt et al. 1995).

The majority of LTR *bel* mRNAs are spliced similarly to those described above for the IP (Fig. 4). However, some contain short, noncoding exons derived from the center of the genome. They are formed with SA sites of the different *env* mRNAs and SD sequences located directly downstream, leading to noncoding exons of about 50 to 272 nt in size (Fig. 4; Muranyi and Flügel 1991; Bodem et al. 1998a).

Several (BFV) or few (FFV) additional splice sites in the *bel* region were detected for which defined protein products cannot be attributed (Renshaw and Casey 1994b; Bodem et al. 1998a). It is unclear yet whether these additional transcripts from the *bel* region have defined functions in FV gene expression or whether they have been detected solely because of the high sensitivity of the reverse transcription-PCR technique used. In analogy to other viral systems, these transcripts are likely to represent by-products of splicing with no intrinsic function.

The data summarized here demonstrate that FV mRNA synthesis and processing are more complex than those known from any other retrovirus.

5
Posttranscriptional Control of FV Gene Expression

Other complex retroviruses like the lentivirus HIV and the HTLV/BLV retroviruses encode posttranscriptional regulators for viral gene expression, which allow nuclear export and cytoplasmic expression of unspliced and partially spliced mRNAs. This gives the virus the capacity to switch from early to late viral gene expression (Cullen 1992, 2000b). A protein functionally corresponding to the HIV Rev or HTLV Rex that achieves these functions has not been identified so far for any of the known FVs (Yu and Linial 1993; Baunach et al. 1993; Lee et al. 1994a;

Adachi et al. 1995; our unpublished data). As described above, deletion of FV Tas abrogates viral gene expression, whereas deletion of the *bel 2*, *bet*, and *bel 3* genes did not result in a phenotype indicative of a post-transcriptional regulator protein (Yu and Linial 1993; Baunach et al. 1993; Lee et al. 1994a; Adachi et al. 1995; Alke et al. 2001). It is, however, possible that FVs contain and utilize a constitutive mRNA transport element (CTE) as known from simpler retroviruses (Bray et al., 1994). Recently, evidence for the existence of a CTE function in the PFV *pol* gene was presented (Wodrich et al. 2001).

The temporal control of early, regulatory, and accessory gene expression versus the late expression of structural genes may be achieved in FVs by the two independent and functionally active transcription units. The promoter in the LTR is primarily used for late structural gene expression, whereas the IP is used to direct *bel* gene expression during all stages of viral gene expression (Löchelt et al. 1995; Bodem et al. 1997; Meiering and Linial 2002).

A functional homolog for the second function of Rev and Rex proteins that allows exit of partially spliced viral transcripts from the nucleus may be the recently reported CTE-like element in PFV *pol* (Wodrich et al. 2001). This is in line with the assumption that a corresponding mechanism should exist in FVs to guarantee that unspliced and partially spliced transcripts encoding structural proteins are efficiently exported from the nucleus. In general, unspliced and even partially spliced viral and cellular mRNAs are actively retained in the nucleus and subsequently degraded (Cullen 2000b). In complex retroviruses, specific RNA sequences have been identified that are responsible for the nuclear retention of transcripts (Schwartz et al. 1992). These sequences are either spliced out or counteracted by nuclear export-promoting regions of the mRNAs, for instance, the constitutive export signal of Mason-Pfizer monkey virus (Bray et al. 1994). Alternatively, the viral Rev and Rex proteins bind defined RNA secondary structures with high specificity and allow the nuclear exit of these transcripts. Distinct components of the different nuclear export pathways, for instance, the cellular TAP and p15 proteins, have been identified (Kang et al. 2000).

For FV LTR-derived structural gene transcripts, extensive splicing in the *env-bel* region has been detected (see above) and can be anticipated for *pol* mRNAs. Besides expressing variant forms of Env, the coding capacity of the partially spliced *gag* and *pol* transcripts is not altered. Thus these splicing events not only extend the coding capacity of FVs but may be primarily utilized as a mechanism to export structural gene transcripts out of the nucleus. Previously, splicing of mRNAs has been considered as a mechanism to directly circumvent nuclear retention (Cullen 2000a). Alternatively, *cis*-inhibitory sequences may be spliced out, but this has not been investigated for FVs.

An additional level of posttranscriptional control of FV gene expression has been described for PFV and SFV-1 (Heinkelein et al. 2000, 2002; Park and Mergia 2000; Russell et al. 2001). Deletion of defined sequences in the PFV R region and thus only contained within *gag*/genomic transcripts have been shown to completely abrogate PFV Gag and Pol protein expression from CMV promoter-driven subgenomic expression constructs (Heinkelein et al. 2000, 2002). The deletions introduced and especially the inactivation of the major SD are likely to have deleted or disrupted essential *cis*-acting motifs on the transcripts because mRNA levels were not altered. It is possible that similar regulatory sequences are located in the long untranslated *pol* mRNA leader or in other FV-specific transcripts, for instance, the CTE-like element identified in PFV *pol* (Wodrich et al. 2001). Other deletions in this region affect proteolytic processing of the Gag protein, pointing to important roles of this RNA sequence in protein expression and processing and also in genomic RNA packaging (Heinkelein et al. 2000). The posttranscriptional control of FV gene expression appears to be complex but is not presently understood.

6
Role of Other Viral Gene Products in the Regulation of FV Gene Expression

The role of the FV Tas protein as the virus-specific transcriptional transactivator is well established. Although Tas-deleted proviruses driven by heterologous promoters substituting for the U3 part of the 5' LTR showed the expression of structural genes, it is an open question whether Tas has additional functions for FV gene expression (Schenk et al. 1999). This seems possible because a multitude of biological activities has been attributed to the HTLV-I Tax protein, a retroviral transactivator of comparable size (Yoshida 1997).

In different FV cell culture systems, the cytoplasmic Bet protein is strongly expressed at levels comparable to the Gag structural protein. Bet is expressed by spliced transcripts from both FV promoters (Figs. 3

and 4). The N-terminal shared domain of Tas that augments transactivation in SFV-1, whereas it is almost dispensable in PFV Tas, is fused to the *bel 2* coding sequence. The SD in *tas* is located 88 (PFV), 31 (FFV), and 34 (BFV) codons downstream of the Tas initiation codon. The splice acceptor (SA) in *bel 2* is far upstream of the Bel 2 domain. Bel 2 is thus fully contained within Bet, which both terminate at the same stop codon (Muranyi and Flügel 1991; Renshaw and Casey 1994b; Bodem et al. 1998a). In BFV, additional spliced transcripts derived from the 5' LTR promoter have been detected by reverse transcription-PCR, which may encode Bel 2- and Bet-related proteins of smaller size (Renshaw and Casey 1994b). The function of these proteins and their expression level are unclear, and these transcripts may solely represent by-products of viral gene expression and splicing.

In PFV, *bel 2* and *bet* can be deleted from full-length infectious molecular clones with no or only minor influence on viral infectivity in cultured cells (Baunach et al. 1993; Yu and Linial 1993). PFV Bel 2 and Bet do not have transactivating capacity when assayed with PFV promoter-based reporter gene constructs, and PFV Bet did not modulate the Tas-mediated transactivation of the PFV LTR-directed gene expression but interfered with the superinfection of Bet-positive cells (Bock et al. 1998). Because Bet can also bind to the PML protein (Regad et al. 2001), it may modulate a PLM-mediated antiviral state. In this context it is worth noting that PFV Bet can shuttle between the cytoplasm and the nucleus of infected cells and that it also shows an intercellular trafficking with the specific cellular release and uptake (Lecellier et al. 2002). How and whether these activities of Bet are related to its suspected function in establishing latency remains to be established (Saïb et al. 1993).

In contrast, data on genetic modifications of the FFV *bel 2/bet* genes that include some amino acid substitutions and even complete genetic knockouts point to an essential and important role of Bet in FFV replication (Alke et al. 2001). Whereas gene expression and processing of viral proteins appeared to be unaffected by the Bet mutation, a reduced release of particles was observed. However, it is likely that additional effects contributed to the strongly reduced viral infectivity (Alke et al. 2001).

7
Foamy Virus-Mediated Alterations of Cellular Gene Expression

One important question to more fully understand FV biology, and one that is directly related to issues of FV-based vector safety and applicability, concerns the interaction of FVs with the infected host cell. Besides interactions that are related to immunity (Winkler et al. 1998; Alke et al. 2000) and direct cytopathogenicity, the interference of FVs with the transcriptional program of the infected cell by direct and indirect mechanisms is very important. The transcriptional activation and repression of cellular genes is especially critical because FVs confer a strong transactivator. The modulation of cellular gene expression has strong implications for the pathogenicity of other complex retroviruses, for instance, HTLV-I (Yoshida 1997).

In a previous report, the activation of the class I major histocompatibility complex promoter by PFV Tas had been described but not followed in more detail (Colas et al. 1995). We undertook a systematic approach to analyze PFV-induced alterations of cellular gene expression with a cDNA array including hundreds of human genes involved in different cellular processes (Wagner et al. 2000).

Distinct cellular genes were reproducibly activated by PFV infection of permissive diploid HEL299 cells: cellular p57Kip2, Egr-1, COUP-TF1, IGF-II, and EphB3 mRNAs were significantly increased in PFV-infected cells and showed a gene-specific and time-dependent induction. The transcriptional activation of these genes was generally concomitant with increased levels of the corresponding proteins. In 293T cells used to characterize the mechanisms of PFV-induced changes in cellular gene expression, the Tas transactivator was necessary and sufficient to strongly increase expression of p57Kip2, IGF-II, and EphB3 genes. In contrast, mRNA levels of COUP-TFI and Egr-1 were only increased in PFV-infected Hel-299 cells but not affected or detectable in Tas-transfected 293T cells (Wagner et al. 2000). This indicates the existence of different, Tas-dependent and Tas-independent, mechanisms of PFV-mediated activation of cellular gene expression and/or a host cell dependence (Fig. 5). Among Tas-independent mechanisms, cell stress and FV-mediated cytotoxicity can be anticipated to play important roles. As outlined in Fig. 5, the activation of cellular genes may directly or indirectly influence not only other cellular but also FV gene expression independent of whether it was initiated in a Tas-dependent or a Tas-independent way.

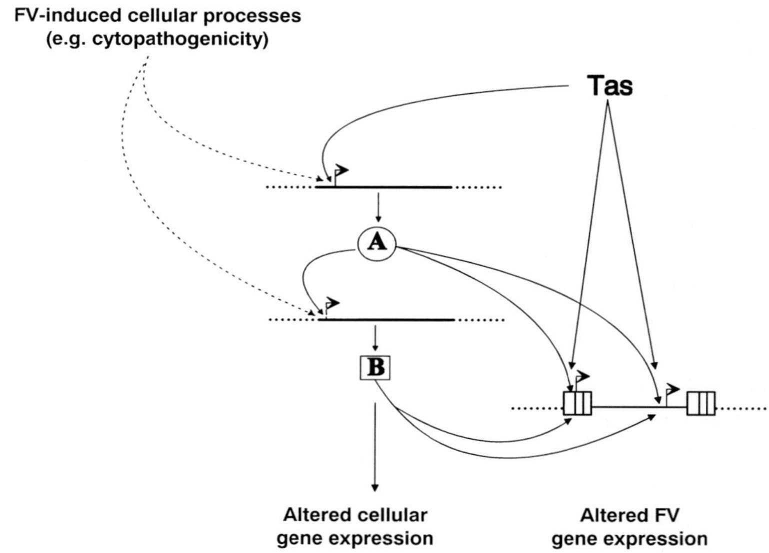

Fig. 5. Hypothetical scheme of direct and/or indirect FV- and Tas-mediated alterations of cellular and viral gene expression. Tas activates FV gene expression (*straight arrows* extending from Tas) and cellular genes (*bent arrows*), which may in turn influence other cellular genes or directly affect FV gene expression. Tas-independent effects of FV infection, e.g., cytopathogenicity, may concomitantly influence cellular gene expression. Promoters are shown as *angular arrows*, the FV LTRs as *open boxes* subdivided into U3-R-U5, and cellular DNA sequences flanking defined genes are shown as *stippled lines*. Hypothetical cellular proteins A and B are shown as *open boxes* and *circles*, respectively. Tas may also repress cellular gene expression, which for clarity is not shown here

Additional cellular genes influenced by PFV might have escaped detection because only late events during PFV infection were studied and the PFV-induced downmodulation of cellular gene expression was not detectable for technical reasons as discussed in that report (Wagner et al. 2000).

Subsequently, the DNA sequences conferring Tas responsiveness to the cellular p57Kip2 gene were shown to be located in the Kip2 coding sequence (Kido et al. 2002). The response element consists of three 14-mer direct repeats that show homology to the TBE of the PFV IP. In fact, Tas specifically bound to the functionally identified cellular TBE with high specificity (Kido et al. 2002). The identification of this cellular TBE may allow subsequent identification of additional Tas-activated cellular genes (Kido et al. 2002).

Upon identification of FV-induced cellular genes, the biological consequences of the increased expression of defined cellular genes for FV replication and gene expression must be analyzed. First data demonstrate that PFV-infected cells release growth-promoting factors, which is in line with our observation of an PFV Tas-mediated activation of IGF-II expression. Furthermore, the molecular mechanisms of Tas-mediated gene expression and the targets of Tas during PFV-induced activation of cellular gene expression are unknown. For instance, Bet may modulate the PFV Tas-mediated activation of cellular genes, which could be one of the critical functions of Bet for FV replication; however, these data must be further extended.

In summary, the data on the Tas-mediated and Tas-independent upregulation of cellular gene expression indicate that a multitude of mutual interactions between FVs and the infected host and host cell take place. From the view of the virus, these interactions may be aimed at enhancing and optimizing viral replication whereas the host (cell) counteracts viral replication by the induction of genes that actively eliminate the virus and/or reduce virus-induced pathology and pathogenicity. This direct interplay may have resulted in the present-day situation that FVs persist efficiently in their host without displaying a detectable degree of pathogenicity. This does not, however, exclude the possibility of FVs playing a role as defined cofactors in different multifactorial disease complexes. In this context, it is important to note that PFV Tas was shown to transactivate HIV LTR-directed gene expression (Keller et al. 1992; Lee et al. 1992). Whether gene expression of other viruses is activated by FVs is unknown.

8
Regulation of LTR- and Internal Promoter-Directed Gene Expression

The regulation of FV gene expression appears to be fundamentally different from that of other complex retroviruses (Cullen 1992). HIV and HTLV, for instance, encode potent posttranscriptional regulators of gene expression allowing the synthesis of structural proteins late during infection after the early phase of regulatory and accessory gene expression. This temporal switch is achieved by the Rev and Rex proteins, which have to accumulate during the initial phase of gene expression.

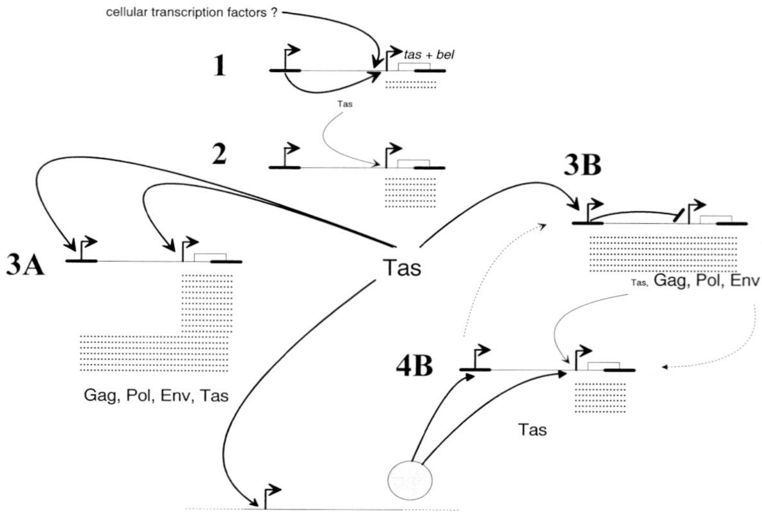

Fig. 6. Schematic diagram of the regulation of FV gene expression. Symbols are as in Fig. 5 except that the FV LTRs are shown as *solid horizontal bars*. Transcripts are shown as *stippled lines* below the coding region of the FV genomes, and their number corresponds to their expression level. Similarly, the size of the names of the different FV proteins is intended to reflect their expression level. A cellular protein affecting both FV promoters is shown as a *filled circle*. Numbers 1 to 4B mark defined phases of FV gene expression (for details, see text)

When a distinct threshold level of these regulatory proteins is reached, the nuclear export and cytoplasmic translation of partially or unspliced structural gene transcripts starts to initiate the late phase of virus replication, which ends with the assembly and release of infectious progeny virus.

FV gene expression is considered to start at the internal promoter as discussed above (Bodem et al. 1997; Fig. 6, phase 1). The switch from an early IP-directed phase of *bel* gene expression (Fig. 6, phases 1 and 2) to a late phase of LTR-driven structural gene expression (phases 3 and 4) may fulfill a function analogous to that of the Rev and Rex proteins. Similar to some DNA viruses, two independent transcription units with different features are activated by the same viral transactivator (Renne et al. 1996). The LTR promoter is absolutely required for structural gene expression, and the biological significance of the IP has been shown experimentally (Löchelt et al. 1995). Different CMV promoter-driven recombinant PFV genomes that lacked or contained the *bel* genes showed strongly reduced viral titers (Schenk et al. 1999). This is in line with the assumption that the temporal and functional interplay of both promoters is crucial for FV infectivity and replication (Bodem et al. 1997).

The data presented above on the activity and regulation of both FV promoters and the Tas transactivator were primarily derived from studies using cells that allow a productive, lytic FV infection. In permissive and semipermissive cells, FV gene expression should be initiated at the IP because of its higher basal activity in the absence of Tas (Löchelt et al. 1993a; Meiering et al. 2001; Meiering and Linial 2002; Fig. 6). Because the IP TRE has the highest affinity to Tas, a relatively low-level Tas expression by the IP may be sufficient to initiate IP-directed Tas-transactivated gene expression (Kang et al. 1998; Fig. 6, phase 2). However, PFV gene expression is strongly repressed by cellular NFI probably through interaction with the IP target sequence (Kido et al 2003). The phase of IP-directed Bel protein expression characterizes the early phase of FV gene expression. When Tas levels increase, the LTR promoter is in turn transactivated by the lower-affinity TBSs, initiating the late phase of structural gene expression. It is unknown whether the switch from early to late gene expression described here is reversible as proposed (Bodem et al. 1997, Fig. 6, stippled arrows). Correspondingly, LTR-directed transcription may suppresses the activity of the IP (Fig. 6, phase 3B). Subsequently Tas expression, which is almost exclusively directed by the IP, is suppressed leading to decreased Tas concentrations. Tas concentrations below a defined threshold level may no longer be sufficient to transactivate the LTR, thus making possible the reinitiation of this regulatory circuit (Bodem et al. 1997, Fig. 6, phase 4B). In line with this, the IP-directed gene expression is more robust under conditions in which the LTR is less active (Meiering et al. 2001) Alternatively, both FV transcription units are active late in infection as displayed in Fig. 6, phase 3A.

It was shown recently that in fully permissive cells, the LTR promoter and the IP are both efficiently transactivated by Tas. In contrast, in cells allowing only a persistent infection, Tas transactivation of the LTR is compromised compared to transactivation of the IP, resulting in the preferential utilization of the IP in persistently infected cells (Linial 2000; Meiering et al. 2001). In the semipermissive cells, phorbol ester treatment increased the basal activity of the IP and significantly enhanced Tas-mediated transactivation of both promoters (Meiering et al. 2001). In subsequent studies it was shown that the intracellular concentration of Tas obtained in the respective cell determines whether a persistent or a lytic replication takes place: only a critical level of Tas expression allows a lytic infection (Meiering and Linial 2002). These data on a certain Tas threshold correspond nicely with data on the action of Rev/Rex and with a model previously discussed for PFV (Bodem et al. 1997).

The establishment of a persistent state in FVs may be regulated by DNA methylation (Schweizer et al. 1993) and histone acetylation (Hatama et al. 2001). Whereas PML appears to be not involved in persistence, Bet has been shown to almost completely abolish the phorbol ester-mediated activation of persisting PFV genomes (Meiering and Linial 2003). These data provide new insights into the establishment of persistence; however, we are far from a detailed understanding of the contributions of viral and cellular factors.

Another open point is whether the Tas-mediated activation of cellular genes, e.g., cellular transcriptional activators, further modulate and influence FV gene expression as shown schematically in Figs. 5 and 6. This is possible because defined transcription factors are upregulated during PFV infection or directly by Tas. However, response elements for these factors have not been identified or functionally characterized in the PFV genome. The upregulated factor may even indirectly modulate FV transcription by the induction or repression of defined signaling cascades that finally affect a cellular mediator that in turn directly interferes with PFV gene expression. Cell type-specific differences in the transactivation or repression of cellular genes by Tas may result in significant alterations in host cell physiology and may thus determine whether a lytic, productive or a semipermissive, persistent infection is established; experiments of this kind should be performed. However, it may be not easy to dissect this intricate network of fine regulation and mutual interference that has been established during a long coevolution between the FVs and their natural hosts.

9
Conclusions

Gene expression of FVs is complex and has unique features in comparison to orthoretroviruses. Whereas the basic features of these FV-specific pathways have been established for distantly related FVs in independent studies, the basic molecular mechanisms, regulatory elements, and functional interrelation of the participating viral and cellular components are only poorly understood. The characterization and molecular definition of these mechanisms will represent a challenge for the future. These studies will probably help us to understand corresponding cellular processes. In this respect, FVs may represent a suitable model for studying the influence of splicing on nucleo-cytoplasmic export and translation. Finally, the interaction of the strong viral transactivator with cellular genes and transcription programs may offer the chance to identify and characterize factors that allow the apparently "peaceful" coexistence of FVs with their natural hosts. This aspect, in particular, will extend our understanding of virus-induced suppression of disease and pathology. Analyzing the persistence and coexistence of FVs in their native hosts thus should allow definition of the underlying molecular mechanisms in an authentic and naturally balanced, undisturbed status.

Acknowledgements. I thank Harald zur Hausen for continuous support and Bryan Cullen, Jennifer Reed, Alexandra Alke, Jochen Bodem, and Rolf M. Flügel for suggestions and for critically reading the manuscript. The work of the author was in part supported by the Deutsche Forschungsgemeinschaft, grant LO 700/1-2.

References

Adachi A, Sakai H, Tokunaga K, Kawamura M (1995) Functional analysis of human spuma retrovirus genome. Virus Genes 11:15–20

Alke A, Schwantes A, Kido K, Flötenmeyer M, Flügel RM, Löchelt M (2001) The bet gene of feline foamy virus is required for virus replication. Virology 287:310–320

Alke A, Schwantes A, Zemba M, Flügel RM, Löchelt M (2000) Characterization of the humoral immune response and virus replication in cats experimentally infected with feline foamy virus. Virology 275:170–176

Baldwin DN, Linial ML (1998) The roles of pol and env in the assembly pathway of human foamy virus. J Virol 72:3658–3665

Baunach G, Maurer B, Hahn H, Kranz M, Rethwilm A (1993) Functional analysis of human foamy virus accessory reading frames. J Virol 67:5411–5418

Blair WS, Bogerd H, Cullen BR (1994) Genetic analysis indicates that the human foamy virus Bel-1 protein contains a transcription activation domain of the acidic class. J Virol 68:3803–3808

Bock M, Heinkelein M, Lindemann D, Rethwilm A (1998) Cells expressing the human foamy virus (HFV) accessory Bet protein are resistant to productive HFV superinfection. Virology 250:194–204

Bodem J, Löchelt M, Delius H, Flügel RM (1998a) Detection of subgenomic cDNAs and mapping of feline foamy virus mRNAs reveals complex patterns of transcription. Virology 244:417–426

Bodem J, Löchelt M, Winkler I, Flower RP, Delius H, Flügel RM (1996) Characterization of the spliced pol transcript of feline foamy virus: the splice acceptor site of the pol transcript is located in gag of foamy viruses. J Virol 70:9024–9027

Bodem J, Löchelt M, Yang P, Flügel RM (1997) Regulation of gene expression by human foamy virus and potentials of foamy viral vectors. Stem Cells 15:141–147

Bodem J, Zemba M, Flügel RM (1998b) Nuclear localization of the functional Bel 1 transactivator but not of the gag proteins of the feline foamy virus. Virology 251:22–27

Bray M, Prasad S, Dubay JW, Hunter E, Jeang KT, Rekosh D, Hammarskjold ML (1994) A small element from the Mason-Pfizer monkey virus genome makes human immunodeficiency virus type 1 expression and replication Rev-independent. Proc Natl Acad Sci U S A 91:1256–1260

Butsch M, Boris-Lawrie K (2002) Destiny of unspliced retroviral RNA: ribosome and/or virion? J Virol 76:3089–3094

Campbell M, Eng C, Luciw PA (1996) The simian foamy virus type 1 transcriptional transactivator (Tas) binds and activates an enhancer element in the gag gene. J Virol 70:6847–6855

Campbell M, Renshaw-Gegg L, Renne R, Luciw PA (1994) Characterization of the internal promoter of simian foamy viruses. J Virol 68:4811–4820

Carter, KC Wang, L Shell, LK Zamir, I Berger, SL Moore PA (1997) The human transcriptional adaptor genes TADA2L and GCN5L2 colocalize to chromosome 17q12-q21 and display a similar tissue expression pattern. Genomics 40:497–500

Chang J, Lee KJ, Jang KL, Lee EK, Baek GH, Sung YC (1995) Human foamy virus Bel1 transactivator contains a bipartite nuclear localization determinant which is sensitive to protein context and triple multimerization domains. J Virol 69:801–808

Coffin, J. M. 1996. Retroviridae and their replication. In Fields, BN Knipe, DM Howley PM (eds) Virology. Raven Press, New York

Colas S, Bourge JF, Wybier J, Chelbi-Alix MK, Paul P, Emanoil-Ravier R (1995) Human foamy virus infection activates class I major histocompatibility complex antigen expression. J Gen Virol 76:661–667

Cullen BR (1991) Human immunodeficiency virus as a prototypic complex retrovirus. J Virol 65:1053–1056

Cullen BR (1992) Mechanism of action of regulatory proteins encoded by complex retroviruses. Microbiol Rev 56:375–394

Cullen BR (2000a) Connections between the processing and nuclear export of mRNA: evidence for an export license? Proc Natl Acad Sci U S A 97:4–6

Cullen BR (2000b) Nuclear RNA export pathways. Mol Cell Biol 20:4181–4187

Enssle J, Jordan I, Mauer B, Rethwilm A (1996) Foamy virus reverse transcriptase is expressed independently from the Gag protein. Proc Natl Acad Sci U S A 93:4137–4141

Erlwein O, Rethwilm A (1993) BEL-1 transactivator responsive sequences in the long terminal repeat of human foamy virus. Virology 196:256–268

Falcone V, Leupold J, Clotten J, Urbanyi E, Herchenröder O, Spatz W, Volk B, Bohm N, Toniolo A, Neumann-Haefelin D, Schweizer M (1999) Sites of simian foamy virus persistence in naturally infected African green monkeys: latent provirus is ubiquitous, whereas viral replication is restricted to the oral mucosa. Virology 257:7–14

Flügel RM, Rethwilm A, Maurer B, Darai G (1987) Nucleotide sequence analysis of the env gene and its flanking regions of the human spumaretrovirus reveals two novel genes. EMBO J 6:2077–2084

Garrett ED, He F, Bogerd HP, Cullen BR (1993) Transcriptional trans activators of human and simian foamy viruses contain a small, highly conserved activation domain. J Virol 67:6824–6827

Giron ML, de The H, Saïb A (1998) An evolutionarily conserved splice generates a secreted env-Bet fusion protein during human foamy virus infection. J Virol 72:4906–4910

Hatama S, Otake K, Ohta M, Kobayashi M, Imakawa K, Ikemoto A, Okuyama H, Mochizuki M, Miyazawa T, Tohya Y, Fujii Y, Takahashi E (2001) Reactivation of feline foamy virus from a chronically infected feline renal cell line by trichostatin A. Virology 283:315–323

He F, Blair WS, Fukushima J, Cullen BR (1996) The human foamy virus Bel-1 transcription factor is a sequence-specific DNA binding protein. J Virol 70:3902–3908

He F, Sun JD, Garrett ED, Cullen BR (1993) Functional organization of the Bel-1 trans activator of human foamy virus. J Virol 67:1896–1904

Heinkelein M, Dressler M, Jarmy G, Rammling M, Imrich H, Thurow J, Lindemann D, Rethwilm A. (2002) Improved primate foamy virus vectors and packaging constructs. J Virol 76:3774–3783

Heinkelein M, Thurow J, Dressler M, Imrich H, Neumann-Haefelin D, McClure MO, Rethwilm A (2000) Complex effects of deletions in the 5' untranslated region of primate foamy virus on viral gene expression and RNA packaging. J Virol 74:3141–3148

Herchenröder O, Turek R, Neumann-Haefelin D, Rethwilm A, Schneider J (1995) Infectious proviral clones of chimpanzee foamy virus (SFVcpz) generated by long PCR reveal close functional relatedness to human foamy virus Virology 214:685–689

Holzschu DL, Delaney MA, Renshaw RW, Casey JW (1998) The nucleotide sequence and spliced pol mRNA levels of the nonprimate spumavirus bovine foamy virus. J Virol 72:2177–2182

Jordan I, Enssle J, Guttler E, Mauer B, Rethwilm A (1996) Expression of human foamy virus reverse transcriptase involves a spliced pol mRNA. Virology 224:314–319

Kang Y, Blair WS, Cullen BR (1998) Identification and functional characterization of a high-affinity Bel-1 DNA binding site located in the human foamy virus internal promoter. J Virol 72:504–511

Kang Y, Bogerd HP, Cullen BR (2000) Analysis of cellular factors that mediate nuclear export of RNAs bearing the mason-pfizer monkey virus constitutive transport element. J Virol 74:5863–5871

Kang Y, Cullen BR (1998) Derivation and functional characterization of a consensus DNA binding sequence for the tas transcriptional activator of simian foamy virus type 1. J Virol 72:5502–5509

Keller A, Garrett ED, Cullen BR (1992) The Bel-1 protein of human foamy virus activates human immunodeficiency virus type 1 gene expression via a novel DNA target site. J Virol 66:3946–3949

Keller A, Partin KM, Löchelt M, Bannert H, Flügel RM, Cullen BR (1991) Characterization of the transcriptional trans activator of human foamy retrovirus. J Virol 65:2589–2594

Kido K, Doerks A, Löchelt M, Flügel RM (2002) Identification and functional characterization of an intragenic DNA binding site for the spumaretroviral trans-activator in the human p57Kip2 gene. J Biol Chem 277:12032–12039

Kido K, Bannert H, Gronostajski RM, Flügel RM (2003) Bel1-mediated trans-activation of the spumaretroviral internal promoter is repressed by nuclear factor I. J Biol Chem (in press)

Lecellier CH, Saïb A (2000) Foamy viruses: between retroviruses and pararetroviruses. Virology 271:1–8

Lecellier CH, Vermeulen W, Bachelerie F, Giron ML, Saïb A. (2002) Intra- and intercellular trafficking of the foamy virus auxiliary bet protein. J Virol 76:3388–3394

Lee AH, Lee KJ, Kim S, Sung YC (1992) Transactivation of human immunodeficiency virus type 1 long terminal repeat-directed gene expression by the human foamy virus bel1 protein requires a specific DNA sequence. J Virol 66:3236–3240

Lee, AH, Lee HY, Sung YC (1994a) The gene expression of human foamy virus does not require a post- transcriptional transactivator. Virology 204:409–413

Lee CW, Chang J, Lee KJ, Sung YC (1994b) The Bel1 protein of human foamy virus contains one positive and two negative control regions which regulate a distinct activation domain of 30 amino acids. J Virol 68:2708–2719

Lee KJ, Lee AH, Sung YC (1993) Multiple positive and negative cis-acting elements that mediate transactivation by bel1 in the long terminal repeat of human foamy virus. J Virol 67:2317–2326

Lindemann D, Rethwilm A (1998) Characterization of a human foamy virus 170-kilodalton Env-Bet fusion protein generated by alternative splicing. J Virol 72:4088–4094

Linial M (2000) Why aren't foamy viruses pathogenic? Trends Microbiol 8: 284–289

Linial ML (1999) Foamy viruses are unconventional retroviruses. J Virol 73:1747–1755

Löchelt M, Aboud M, Flügel RM (1993a) Increase in the basal transcriptional activity of the human foamy virus internal promoter by the homologous long terminal repeat promoter in cis. Nucleic Acids Res 21:4226–4230

Löchelt M, Flügel RM (1995) The molecular biology of primate spumaviruses. In Levy JA (ed.) The Retroviridae vol. 4. Plenum Press, New York

Löchelt M, Flügel RM (1996) The human foamy virus pol gene is expressed as a Pro-Pol polyprotein and not as a Gag-Pol fusion protein. J Virol 70:1033–1040

Löchelt M, Flügel RM, Aboud M (1994) The human foamy virus internal promoter directs the expression of the functional Bel 1 transactivator and Bet protein early after infection. J Virol 68:638–645

Löchelt M, Muranyi W, Flügel RM (1993b) Human foamy virus genome possesses an internal, Bel-1-dependent and functional promoter. Proc Natl Acad Sci U S A 90:7317–7321

Löchelt M, Yu SF, Linial ML, Flügel RM (1995) The human foamy virus internal promoter is required for efficient gene expression and infectivity. Virology 206:601–610

Löchelt M, Zentgraf H, Flügel RM (1991) Construction of an infectious DNA clone of the full-length human spumaretrovirus genome and mutagenesis of the bel 1 gene. Virology 184:43–54
Maurer B, Serfling E, ter Meulen V, Rethwilm A (1991) Transcription factor AP-1 modulates the activity of the human foamy virus long terminal repeat. J Virol 65:6353–6357
Meiering CD, Rubio C, May C, Linial ML (2001) Cell-type-specific regulation of the two foamy virus promoters. J Virol 75:6547–6557
Meiering CD, Linial ML (2001) Historical perspective of foamy virus epidemiology and infection. Clin Microbiol Rev 14:165–176
Meiering CD, Linial ML (2002) Reactivation of a complex retrovirus is controlled by a molecular switch and is inhibited by a viral protein. Proc Natl Acad Sci USA 99:15130–15135
Meiering CD, Linial ML (2003) The promyelocytic leukemia protein does not mediate foamyvirus latency in vitro. J Virol 77:2207–2213
Mergia A (1994) Simian foamy virus type 1 contains a second promoter located at the 3' end of the env gene. Virology 199:219–222
Mergia A, Pratt-Lowe E, Shaw KE, Renshaw-Gegg LW, Luciw PA (1992) cis-acting regulatory regions in the long terminal repeat of simian foamy virus type 1. J Virol 66:251–257
Mergia A, Renshaw-Gegg LW, Stout MW, Renne R, Herchenroeder O (1993) Functional domains of the simian foamy virus type 1 transcriptional transactivator (Taf). J Virol 67:4598–4604
Mergia A, Shaw KE, Pratt-Lowe E, Barry PA, Luciw PA (1990) Simian foamy virus type 1 is a retrovirus which encodes a transcriptional transactivator. J Virol 64:3598–3604
Mergia A, Shaw KE, Pratt-Lowe E, Barry PA, Luciw PA (1991) Identification of the simian foamy virus transcriptional transactivator gene (taf). J Virol 65:2903–2909
Moebes A, Enssle J, Bieniasz PD, Heinkelein M, Lindemann D, Bock M, McClure MO, Rethwilm A (1997) Human foamy virus reverse transcription that occurs late in the viral replication cycle. J Virol 71:7305–7311
Muranyi W, Flügel RM (1991) Analysis of splicing patterns of human spumaretrovirus by polymerase chain reaction reveals complex RNA structures. J Virol 65:727–735
Park J, Mergia A (2000) Mutational analysis of the 5' leader region of simian foamy virus type 1. Virology 274:203–212.
Regad T, Saïb A, Lallemand-Breitenbach V, Pandolfi PP, de The H, Chelbi-Alix MK (2001) PML mediates the interferon-induced antiviral state against a complex retrovirus via its association with the viral transactivator. EMBO J 20:3495–3505
Renne R, Fleps U, Luciw PA, Neumann-Haefelin D (1996) Transactivation of the two promoters of SFV-3 by different mechanisms. Virology 221:362–367
Renne R, Mergia A, Renshaw-Gegg LW, Neumann-Haefelin D, Luciw PA (1993) Regulatory elements in the long terminal repeat (LTR) of simian foamy virus type 3 (SFV-3). Virology 192:365–369
Renshaw RW, Casey JW (1994a) Analysis of the 5' long terminal repeat of bovine syncytial virus. Gene 141:221–224

Renshaw RW, Casey JW (1994b) Transcriptional mapping of the 3' end of the bovine syncytial virus genome. J Virol 68:1021–1028
Rethwilm A (1995) Regulation of foamy virus gene expression. Curr Top Microbiol Immunol 193:1–24
Rethwilm A, Erlwein O, Baunach G, Maurer B, ter Meulen V (1991) The transcriptional transactivator of human foamy virus maps to the bel 1 genomic region. Proc Natl Acad Sci U S A 88:941–945
Russell RA, Zeng Y, Erlwein O, Cullen BR, McClure MO (2001) The R region found in the human foamy virus long terminal repeat is critical for both Gag and Pol protein expression. J Virol 75:6817–6824
Saïb A, Neves M, Giron ML, Guillemin MC, Valla J, Peries J, Canivet M (1997) Long-term persistent infection of domestic rabbits by the human foamy virus. Virology 228:263–268
Saïb A, Peries J, de The H (1993) A defective human foamy provirus generated by pregenome splicing. EMBO J 12:4439–4444
Schenk T, Enssle J, Fischer N, Rethwilm A (1999) Replication of a foamy virus mutant with a constitutively active U3 promoter and deleted accessory genes. J Gen Virol 80:1591–1598
Schliephake AW, Rethwilm A (1994) Nuclear localization of foamy virus Gag precursor protein. J Virol 68:4946–4954
Schmidt M, Herchenröder O, Heeney J, Rethwilm A (1997) Long terminal repeat U3 length polymorphism of human foamy virus. Virology 230:167–178
Schwantes A, Ortlepp I, Löchelt M (2002) Construction and functional characterization of feline foamy virus-based retroviral vectors. Virology 301:53–63
Schwartz S, Felber BK, Pavlakis GN (1992) Distinct RNA sequences in the gag region of human immunodeficiency virus type 1 decrease RNA stability and inhibit expression in the absence of Rev protein. J Virol 66:150–159
Schweizer M, Fleps U, Jackle A, Renne R, Turek R, Neumann-Haefelin D (1993) Simian foamy virus type 3 (SFV-3) in latently infected Vero cells: reactivation by demethylation of proviral DNA. Virology 192:663–666
Tobaly-Tapiero J, Bittoun P, Neves M, Guillemin MC, Lecellier CH, Puvion-Dutilleul F, Gicquel B, Zientara S, Giron ML, de The H, Saïb A (2000) Isolation and characterization of an equine foamy virus. J Virol 74:4064–4073
Venkatesh, LK, Chinnadurai G (1993) The carboxy-terminal transcription enhancement region of the human spumaretrovirus transactivator contains discrete determinants of the activator function. J Virol 67:3868–3876.
Venkatesh LK, Theodorakis PA, Chinnadurai G (1991) Distinct cis-acting regions in U3 regulate trans-activation of the human spumaretrovirus long terminal repeat by the viral bel1 gene product. Nucleic Acids Res 19:3661–3666
Venkatesh LK, Yang C, Theodorakis PA, Chinnadurai G (1993) Functional dissection of the human spumaretrovirus transactivator identifies distinct classes of dominant-negative mutants. J Virol 67:161–169
Wagner A, Doerks A, Aboud M, Alonso A, Tokino T, Flügel RM, Löchelt M (2000) Induction of cellular genes is mediated by the Bel1 transactivator in foamy virus-infected human cells. J Virol 74:4441–4447
Winkler I, Bodem J, Haas L, Zemba M, Delius H, Flower R, Flügel RM, Löchelt M (1997) Characterization of the genome of feline foamy virus and its proteins

shows distinct features different from those of primate spumaviruses. J Virol 71:6727–6741
Winkler IG, Flügel RM, Löchelt M, Flower RL (1998) Detection and molecular characterisation of feline foamy virus serotypes in naturally infected cats. Virology 247:144–151
Winkler IG, Löchelt M, Flower RL (1999) Epidemiology of feline foamy virus and feline immunodeficiency virus infections in domestic and feral cats: a seroepidemiological study. J Clin Microbiol 37:2848–2851
Wodrich H, Bohne J, Gumz E, Welker R, Krausslich HG. (2001) A new RNA element located in the coding region of a murine endogenous retrovirus can functionally replace the Rev/Rev-responsive element system in human immunodeficiency virus type 1 Gag expression. J. Virol 75:10670–10682
Yang P, Zemba M, Aboud M, Flügel RM, Löchelt M (1997) Deletion analysis of both the long terminal repeat and the internal promoters of the human foamy virus. Virus Genes 15:17–23
Yoshida M (1997) Howard Temin memorial lectureship. Molecular biology of HTLV-1: deregulation of host cell gene expression and cell cycle. Leukemia 11:14–15
Yu SF, Baldwin DN, Gwynn SR, Yendapalli S, Linial ML (1996a) Human foamy virus replication: a pathway distinct from that of retroviruses and hepadnaviruses. Science 271:1579–1582
Yu SF, Edelmann K, Strong RK, Moebes A, Rethwilm A, Linial ML (1996b) The carboxyl terminus of the human foamy virus Gag protein contains separable nucleic acid binding and nuclear transport domains. J Virol 70:8255–8262
Yu SF, Linial ML (1993) Analysis of the role of the bel and bet open reading frames of human foamy virus by using a new quantitative assay. J Virol 67:6618–6624
Zou JX, Luciw PA (1996) The transcriptional transactivator of simian foamy virus 1 binds to a DNA target element in the viral internal promoter. Proc Natl Acad Sci USA 93:326–330

Chapter 3

Proteolytic Processing of Foamy Virus Gag and Pol Proteins

R. M. Flügel, K.-I. Pfrepper

Abstract The foamy viral proteases (FV PRs) are set apart from other retroviral processing enzymes by unique features. The first remarkable property is that FV PRs are enzymatically active as high-molecular-mass Pro-Pol proteins. Hence there exist multiple forms of active FV PRs that likely contribute to cleavage site specificity. A FV PR of low molecular

R. M. Flügel (✉)
Retroviral Gene Expression, Research Programme Applied Tumor Virology, German Cancer Research Center,
Im Neuenheimer Feld 242, 69009 Heidelberg, Germany
E-mail: r.m fluegel@dkfz-heidelberg.de

K.-I. Pfrepper
Institut für Immunologie der Universität Heidelberg,
Universität Heidelberg, Im Neuenheimer Feld 305, 69009 Heidelberg, Germany

size is not detectable in purified virions, in contrast to PRs of other retroviruses that are found in virus particles. Because the major part of Pol remains attached to the amino-terminal enzymatically active PR protein region, the FV-specific way of expressing Pro-Pol polyproteins from a *pol*-specific transcript provides for the incorporation of Pro-Pol and IN into virus particles. Proteolytic processing of Gag and Pol proteins is incomplete and delayed. Another novel feature is that the catalytic center of the active dimers of cat FV PR consists of D-S/T-Q instead of D-S/T-G, an unprecedented feature of this enzyme. The temporal and spatial control and the factors that regulate FV PRs remain to be elucidated.

1 Introduction and Overview on Retroviral and Cellular Aspartic Proteases

Proteolytic processing at specific sites of immature viral precursor proteins by the viral protease (PR) is an obligate step in the life cycle of retroviruses (RVs). Thus, without an enzymatically active viral PR, infectious virions will not be produced. This holds true not only for RVs but also for a wide variety of diverse RNA and DNA viruses (Babe and Craik 1997). Because most viruses are pathogenic, the corresponding PRs provide key targets in the rational design of antiviral agents. The resulting great interest in RV PRs, and particularly, in human immunodeficiency virus (HIV-1) yielded rich information on the structure and function of these enzymes.

The RV PR was detected as a virion-associated enzyme that had the specificity to cleave avian retroviral Gag precursors, forming the mature Gag proteins that had been previously identified (Vogt and Eisenman 1973). Subsequently, the *pro* gene was mapped to a genomic position upstream of the reverse transcriptase (RT) domain of *pol* and downstream of *gag*. The known foamy virus (FV) genomes contain the *pro* gene that codes for PR in the *pol* reading frame. Other RV PRs occur in different reading frames relative to *gag* and *pol* (Vogt 1997). RV PRs have been characterized in great detail by biochemical, genetic, and three-dimensional structural analyses. These studies led to a deeper insight into the structural, functional, and regulatory properties of PRs of RVs that drive the proteolytic processing of the immature Gag and Pol precursor proteins into the mature forms that are required for the completion of the virus life cycle. RV PRs belong to the large aspartic subfamily of proteinases with the conserved D-S/T-G-A motif that is the essential part of the catalytic center.

To be enzymatically active, the viral PR protein has to dimerize (Vogt 1997). These enzymes catalyze peptide bond hydrolysis through an acid-base mechanism mediated by the two conserved catalytic aspartic acid residues of the symmetric homodimer.

Nature has designed alternative ways to form active aspartic PRs, as illustrated by the cellular aspartic PRs renin, pepsin, cathepsin D, and many others. Genetic and phylogenetic evidence indicates that these enzymes evolved by gene duplication of an ancestral gene. In these cases, one long polypeptide chain of up to about 370 amino acid residues (aar) of two highly homologous duplicated units folds into an active PR of relatively large molecular size. These large proteins are initially expressed as preproenzymes (zymogens) that on acidification, and step by step, are autocatalytically cleaved into the mature active enzymes (Khan et al. 1997). The catalytic centers and flanking sequences of both monomers form substrate-binding pockets. The small peptides that are removed during autocatalysis regulate the activity of the mature enzymatic forms (Khan et al. 1997; Dunn 1997; Bernstein et al. 1999). The three-dimensional structures of two intermediates of cathepsin D have been reported, providing detailed insight into how substrate specificity is modulated by intracellular pH (Lee at al. 1998; Dunn et al. 1998). A novel membrane-anchored aspartic PR, called β-secretase, was discovered, which is responsible for proteolysis of the amyloid precursor protein in Alzheimer's disease (Yan et al. 1999). Surprisingly, presenilin that is also involved in Alzheimer's disease has been shown to be identical to γ-secretase, another novel member of a new class of eukaryotic aspartic PRs that has no homology to known aspartic PRs and functions as a heterodimer (Wolfe et al. 1999). In addition, these new cellular PRs have a different cleavage site specificity. The specificity of viral PRs is discussed below.

In contrast to the cellular aspartic PRs, most RV enzymes are small in size and expressed as monomers and, on reaching a critical concentration dimerize spontaneously under proper conditions. They are of similar lengths, ranging from 97 to about 130 aar ,with the notable exception of human foamy virus, SFVcpz(hu) (here abbreviated PFV) Pro-Pol as discussed below.

```
rsv   LAMTMEHKDRPLVRVILTNTGSHPVKQRSVYITALLDSGADITIISEEDWPTDWPV-MEAA          60
      βββββ     ββββββββββ          βββββββββ      βββββ        β βββ

hiv-1 PQITLW--QRPLVTIRIG----------GQLKEALLDTGADDTVLEE--MNLP-GKWKPKM          46
      ββββ      βββββββββ            βββββββββ    ββββββ        ββββ

pfv   MNPLQL--LQPLPAEIK-----------GTKLLAHWDSGATITCIPESFLEDEQPIKKTLI          48
         αα  ααααααα                ββββββ      βββββββ          ββββ


rsv   NPQIHGIGGGIPMRKSRDMIELGVINRDGSLERPLLLFPAVA---MVRGSILGRDCLQGLGLRLTNL•  124
                ββββββββββ      βββββββββββββββ       ββββαααααααααββββ

hiv-1 ---IGGIGGFIKVRQY-DQIPVEIC-----GH-KAIGTVLVG---PTPVNIIGRNLLTQIGCTL-NF•    99
         βββ   βββββββββ ββββββββ       β ββββββββββ      βββαααααααααββββ β

pfv   -KTIHGEKQQNVYYVT----FKVKGRKVE---------AEVIASPYEYILLSPTDVPWLTQQPL-QL   100
       ββββ    βββββββ    ββββββββββ         ββββ    ββββββ     αααα   ββ


pfv   TILVPLQEYQEKILSKTALPEDQKQQLKTLFLKYDNLWQHWENQVGHR......              751 (596)
      βββ  αααααααααα   ααααααααααααααααα
```

Fig. 1. Structural alignment of the three-dimensional structures of RSV and HIV-1 PR domains. The secondary structure for PFV PR was derived by combining the data for the RSV and HIV-1 PRs with the predictions from the phD and DSC programs (Rost and Sander 1994). Underlined residues correspond to the flap sites; the aspartic acid of the catalytic center and the conserved I51 that are essential for dimerization are in *boldface letters*. The *dots* at the carboxyl terminus of the PFV PR indicate that this sequence continues into the RT and RNase H domains; the *numbers* 751 and 596 indicate the last residue of the RT and RNase H domains, respectively. The *thick dot* at the end of RSV and HIV-1 PRs symbolizes the experimentally determined carboxyl termini. Gaps between RSV and HIV-1 PR sequence were introduced to maximize the structural alignment as reported by Wlodawer et al. (1995)

The three-dimensional structure of the symmetric HIV-1 PR homodimer serves as a reference structure for discussions of FV PRs. It reveals that the interface between the two identical subunits is composed of several well-ordered β-sheets shown schematically in Fig. 1. In addition, the 99-aar-long HIV-1 enzyme has residues 84 to 94 in an α-helical conformation. The aspartic residues of both monomers contribute to the active site cleft. The walls or flaps of the cleft lie above the active site and consist of a short antiparallel β-sheet with a turn. The flaps are important factors in determining the substrate specificity of the PRs. Additional determinants of substrate specificity turned out to be side chains of distinct residues further downstream of the flap regions. Another noteworthy feature of the X-ray analyses of several retroviral protease-substrate (or -inhibitor) complexes are the tips of the flaps. They are highly flexible hinges or joints, because NMR analysis of PR-inhibitor complexes showed that the flaps move on binding the inhibitor. Distinct residues

```
        1                                               50
pfv     MNPLQLLQPLPAEIKGTKLLAHWDSGATITCIPESFLEDEQFIKKTLIKT
sfv1    MDPLQLLQPLEAEIKGTKLKAHWDSGATITCVPEAFLEDERFIQTMLIKT
sfv3    MDPLQLLQPLEAEIKGTKLKAHWDSGATITCVPQAFLEEEVFIKNIWIKT
bfv     ...MPALRPLQVEIKGNHLKGYWDSGAEITCVPAIYIIEEQFVGKKLITT
efv     ...MQALQPLQVQIKGNSLKGFYDTGAEITCVPAIFLIEEEFIGERTIQT
ffv     ...MDLLKPLIVERKGVKIKGYWDSQADITCVPKSLLQGEEFVRQQNVTT
              * **  * **       *  * *** *      * *       *

       51                                              100
pfv     IHGEKQQNVYYVTFKVKGRKVEAEVIASPYEYILLSPTDVPWLTQQPLQL
sfv1    IHGEKQQDVYYLTFKVQGRKVEAEVLASPYDYILLNPSDVPWLMKKPLQL
sfv3    IHGEKEQPVYYLTFKIQGRKVEAEVISSPYDYILVSPSDIPWLMKKPLQL
bfv     IHNEKEHDVYYVEMKIEKRKVQCEVIATALDYVLVAPVDIPWYKPGPLEL
efv     IHGITKEKVYYLTFKIQGRKLAAEVIGTQLDYVIIAPSDIPWYKKYELEL
ffv     IHGTQEGDVYYVNLKIDGRRINTEVIGTTLDYAIITPGDVPWILKKPLEL
        **      ***   *   *    **      *    * * **    ** *

      101                                            148
pfv     TILVPLQEYQEKILSKTALPEDQKQQLKTLFVKYDNLWQHWENQVGHR....
sfv1    TVLVPLHEYQEFLLQQTALPKEQKELLQKLFLKYDALWQHWENQVGHR....
sfv3    TTLVPLQEYEEFLLKQTMLTGSYKEKLQSLFLKYDALWQHWENQVGHR....
bfv     TIKIDVESQKHILITESTLSPQGQMRLKKLLDQYQALWQCWENQVGHR....
efv     TIKIDIQKQQECLLHTTNLSSEGKKYLKDLFIKYDNLWQKWENQVGHR....
ffv     TIKLDLEEQQGILLNNSILSKKGKEELKQLFDKYNALWQSWENQVGHR....
        *                 *       *  *   *  *** ********
```

Fig. 2. Alignment of the protease domains of the known foamy viral Pro-Pol polyproteins. The aspartic acid residues are in *boldface letters*. *Asterisks* mark conserved residues conserved in all shown PR sequences. The Q23 residue of the active FFV PR is in *italics* and *underlined* because of its unique nature at this position that is different from other retroviral PRs. E53 and E142 of PFV PR are in *boldface* and in *italics* as these residues have been analyzed by mutagenesis (see text). The *dots* at the carboxyl termini symbolize the fact that FV PRs are active and detectable as long Pol polyproteins (for details, see text)

essential for dimerization are located in the amino-, central, and carboxy-terminal regions (Wlodawer and Gustchina 2000). By and large, the FV PRs are comparable to other RV enzymes in terms of homology. Multiple alignments of PR sequences from one of the group of RVs illustrate the relatively low degree of homology between even closely related RV members. This holds also true for FV PR sequences with a degree of homology of 44% between those of PFV (prototype foamy virus) and feline foamy virus (FFV) (Fig. 2), whereas the other Pol domains possess a higher degree of homology (Löchelt and Flügel 1995). Phylogenetic analysis shows that PR sequences seem to occur ubiquitously in retroelements from yeast to those of *Drosophila* and human. A notable exception is hepatitis B virus (HBV), which does not code for any PR, although the closely related cauliflower mosaic virus encodes an active PR (Coffin et al. 1997).

This review focuses on the properties particularly of the PFV and FFV enzymes. The structural and functional aspects are treated, with emphasis on the special features of FV Pro-Pol proteins that set them apart from other viral PRs. The major focus is on the processing of the Gag and Pol precursor proteins as well as the use of peptides as substrates for assaying protease activity. Comparative aspects of RV PRs are briefly mentioned, and the reader is referred to reviews that have been published elsewhere (Vogt 1997; Wlodawer and Erickson 1993). Although some questions of RV PRs have been partially answered through efforts that resulted in better understanding of the structural determinants of substrate specificity, some issues such as the temporal and spatial regulation of the viral PR activity during the different steps of processing of viral proteins and its significance for disassembly, maturation, and morphogenesis remain unresolved.

2
Biosynthesis and Expression of Enzymatically Active Foamy Viral Pro-Pol Proteins

As most studies have been carried out with either PFV or FFV PR protein domains, the following chapters will deal with these enzymes. To understand the unique way of PR expression in spumaviruses, it is essential to examine closely those of other RVs. In general, formation of the monomeric RV PRs can occur by different pathways that depend on how the corresponding transcripts are expressed: (1) by autocatalytic processing of a large Gag-Pro-Pol fusion protein that is generated through one ribosomal frameshift event at a specific site in the *gag-pro-pol* overlap region for lentiviral PRs; (2) by autocatalytic processing of Gag-Pro-Pol, Gag-Pro, and Pro-Pol polyproteins that are generated through two ribosomal frameshift events at two specific sites in the *gag-pro-pol* overlap regions of mouse mammary tumor virus, Mason-Pfizer monkey virus, and human T-cell lymphotropic virus; (3) in Rous sarcoma virus (RSV), the PR is encoded by *gag*; (4) for murine leukemia virus, the Gag-Pro-Pol polyprotein is generated by amber codon suppression that is located between *gag* and the *pro-pol* reading frame and decoded as Gln. The salient point is that active PR proteins are expressed in multiple forms, and in most viruses, a PR of low molecular size (of about 12–19 kDa) is detectable in purified virions as final reaction product. Most biochemical studies have been performed using this low-molecular-size form that was bacterially expressed and purified, because it is easier to handle. However, a number of questions were not answered by this approach. What is the first cleavage event? Do the longer Gag-Propol and Propol polyproteins contribute to proteolysis at specific sites? Does one PR form cut one or several defined cleavage sites? How are the PR activities regulated? Do they cleave in *cis* and/or in *trans* at either end of the PR protein region of the protease-containing fusion proteins? Because, in contrast to most RVs, FV PR domains have evolved in a way that avoids frameshifting, some of the above-mentioned questions can be approached in a different way as FV Pro-Pol polyproteins are expressed from a spliced *pol*-specific transcript without forming a Gag-Pro-Pol protein (Yu et al. 1996; Enssle et al. 1996; Löchelt and Flügel 1996; Bodem et al. 1996) and without forming a FV PR of small size (Pfrepper et al. 2000).

3
Proteases of Foamy Viruses

There are several remarkable properties of FV PRs that set them apart from proteinases of other RVs. These characteristic features encompass both structural and functional properties. At first sight, FV PRs seem to be similar to other retroviral proteases with respect to location and enzymatic properties. Alignments clearly show that the PR domain of the known FVs is located and encoded at the amino-terminal region of *pol*. The PFV PR domain starts at the first Met of the *pol* reading frame (Löchelt and Flügel 1996).

In most studies of FV PRs, recombinant proteins have been used that were obtained by bacterial expression and subsequently purified by affinity column chromatography. Proteolysis of viral FV proteins by recombinant FV PRs was not due to contaminating bacterial PRs because the inactive PFV PR D/A mutant was inactive (Konvalinka et al. 1995).

An essential requirement for PFV PR activity is that only homodimeric PRs even when bacterially expressed are proteolytically active (Pfrepper et al. 1998). This feature is shared with all other known PRs of RVs.

Discussing first the influence of mutation at the amino-terminal part of the PFV PR domain, it is noteworthy that a recombinant clone Δp3-ProPol that is derived from the infectious pHSRV13 provirus was not capable of cleaving the RNase H-Integrase scissile bond. Because in this mutant proviral clone, the first Met of Pol was replaced by a Thr residue, this result indicates that the inherent proteolytic activity was not functional and relative sensitive to changes at the amino terminus of the PFV PR domain (Löchelt and Flügel 1996).

While analyzing recombinant PFV PR proteins of short size, at first sight, the situation at the putative carboxyl terminus of the PR domain seemed to display similar characteristics. But surprisingly, a carboxyl-terminal processing site of wild-type FV PR sequence has not been observed and does not seem to occur in vivo. Instead, long and active PRRT, PR-RT-RNase H, and PR-RT-RNase H-IN proteins of substantial size are expressed in infected cells (Pfrepper et al. 1999). Furthermore, these long proteins are clearly detectable in virus particles by Western blots and, importantly, possess specific proteolytic activity (Pfrepper et al. 1999). But repeated attempts to detect short PFV PR forms in purified virus particles comparable in size to other RV PRs failed (Pfrepper 1999). Thus the fact is that a native and short PFV and FFV PR protein corresponding in length to other active RV PRs was not detectable in several different cell types under the conditions used. This is in contrast to what has been stated in recent reviews (Linial 1999; Lecellier and Saïb 2000). Partly, this confusion was due to the fact that short recombinant PFV PR proteins of intermediate length (of up to 223 aar) but not of lengths corresponding to the native Pro-Pol proteins autocatalytically cleave themselves when expressed in *Escherichia coli* at the sequence HWEN↓QVGHR (Pfrepper et al. 1998). This cleavage site that is located between N143 and Q144 is cut at a low efficiency. This fostered the misleading suspicion that N143 might be the carboxyl terminus of a short native FV PR. For further discussion of this site in synthetic peptides, see below under specificity of FV PRs. One group reported a 10-kDa protein in PFV-infected cells that reacted with a PR domain-specific antiserum (Netzer et al. 1993). This report remains to be confirmed.

Recombinant PR proteins that are longer than 300 residues seem to block proteolysis at site N143/Q144 by occluding this site (Pfrepper et al. 2000). Comparable observations have been made with the RSV Gag-Pol proteins that are not cleaved between the PR and the RT domain after in vitro assay with RSV PR (Vogt 1997). This is remarkable, because it is the only cleavage site resistant to cleavage in vitro; this also holds true for the PFV cleavage site N143/144 between the PR and RT domains.

Recombinant PFV PR forms were reported to require an acidic pH between 6.0 and 6.6 (Pfrepper et al. 1999; Fenyövalvi et al. 1999). This value is slightly higher than that of other retroviral enzymes and might indicate that some cleavage events are delayed and occur in the infected cells (see below). The recombinant PFV PR also requires a high salt concentration of 2.0 M NaCl. Additionally, synthetic peptides have been used as substrates for RV (Kotler et al. 1988) and FV PRs (Pfrepper et al. 1997). In general, peptide substrates offer advantages compared with recombinant viral proteins, e.g., they are easy to obtain and purify, and they allow a systematic analysis of length requirements and the determination of specific residues at distinct positions within the cleavage site junction.

As another surprise, the FFV PR domain sequence revealed that Gln replaced the ubiquitously occurring Gly residue at the conserved D-S/T-G-A motif, another hint at Nature's capability to vary apparently invariant motifs. To rule out sequencing errors, nucleotides of both strands of the cloned FFV DNA genome and even from different FFV field isolates were determined (Winkler et al. 1997). As further proof that the FFV PR actually contains the motif D-S-Q-A, recombinant forms of the FFV PR protein containing this Gln residue were subsequently shown to be active (see below, under Gag processing).

An alignment of the known FV PR domain sequences clearly reveals the close phylogenetic relatedness (Fig. 2). The comparison of the FV PR domains show that they have the catalytic center motif D-S-Q-A comparable to the D-T-G-A motif of other retroviral PRs; except for RSV PR, which also has the D-S-G-A motif. Conversion of the catalytic center aspartic acid into an alanine residue by site-directed mutagenesis resulted in an inactive PFV PR enzyme as expected (Konvalinka et al. 1995). Mutagenesis of the Ser residue to Thr did not lead to any significant change in enzymatic activity. These studies were carried out in the context of an infectious PFV clone and showed that the active proteolytic activity is absolutely required for viral replication. The recently reported equine foamy virus (EFV) PR sequence contains a Thr residue at this position indicating that this position can be occupied by either Ser or Thr in FV PRs (Tobaly-Tapiero et al. 2000).

The region of the flaps of lentiviral PR is not easily identified in FV PRs. However, neural network programs are helpful to predict FV flaps with secondary structures analogous to those found in other retroviral

Table 1 Locations of proteolytic cleavage sites of spumaviral Gag and Pol proteins

	Gag	Gag	Gag	Gag	Pol	Pol
Virus	p33–p34[b]	p3.5–p1.5[c]	p1.5–p29[c]	p68–p3[a]	PR-RT–RH[b]	PR-RT-RH–IN[a]
HFV	QHIR↓SVTG[d]	DGVF↓PVTT[d]	RIIN↓AILG[d]	RAVN↓TVTQ[d]	EGVF↓YTDG[d]	YVVN↓CNTK[d]
SFV-1	QHIR↓AVTG	EGVF↓PIPTS	RVIN↓ALLG	RSVN↓TVTA	AMVF↓YTDG	YVVH↓CNTT
SFV-3	QHIR↓AVTG	EGVF↓MTTP	RVVN↓ALIG	RNVD↓TVTA	SMVF↓YTDG	YVVN↓INTT
BFV	THIR↓AVIG	QGVY↓PVQD	RTVN↓ALTV	SAVH↓SVRL	EAIF↓YTDG	YKVF↓SNEL
FFV	NHLR↓SVIG	EGVF↓PIVD	RRVN↓ALVA	AAVH↓TVKA[d]	QHIF↓YTDG	FKVH↓MTKN
EFV	THIR↓AVIG	TGVY↓PAMD	RLVN↓AITA	NTVH↓TVRQ	EVVF↓YTDG	YTVF↓SVQT
Cons.	XHIR↓AVTG	XGVF↓PXXX	RXVN↓ALXX	XXVX↓TVXX	XXVF↓YTDG	YXVX↓XXXX

[a] Primary cleavage sites
[b] Secondary cleavage sites
[c] Tertiary cleavage sites
[d] Cleavage sites experimentally determined by N-terminal sequencing or mass spectrometry of the corresponding peptides cleaved

PRs. On the other hand, these programs indicate that there are differences in the secondary structures of FV PR domains that will likely result in three-dimensional structures at least partially different from those of known PRs (Fig. 1).

In the following discussion, Gag and Pol processing are presented and analyzed in close context to outline those features that are shared by both Gag and Pol proteins as substrates for the FV PRs.

4
Proteolytic Processing of Gag and Pol Proteins

4.1
Gag Processing

Table 1 shows the locations of scissile peptide sites that were experimentally determined after assay with recombinant PFV and FFV PRs (Pfrepper et al. 1997, 1998, 1999). The other cleavage sites included in the list were deduced from indirect evidence, i.e., from the published sizes of polypeptide bands after analysis by gel electrophoresis under denaturing conditions. Alternatively, the locations of the Gag cleavage sites were derived by analogy with the known PFV scissile bonds as is the case for SFV types 1 and 3, FFV, EFV, and BFV (Tobaly-Tapiero et al. 1990; Winkler et al. 1997; Tobaly-Tapiero et al. 2000; Holzschu et al. 1998). The identification of the PFV-1 Gag p68/p3 cleavage site (Pfrepper et al. 1997) has been confirmed by Fenyövalvi et al. (1999).

The active FFV PR contains the D-S-Q-A motif and is of special interest because it seems to break the rule of the invariant D-T/S-G-A motif completely conserved in other PRs of RV. To illustrate the approach, we here present data on the corresponding but unpublished FFV Gag p48/p3 scissile bond. As shown in the Western blot of Fig. 3, a purified recombinant FFV Gag protein of about 60 kDa containing a His tag at the carboxyl terminus was cleaved to yield two reaction products (marked by arrows). To increase solubility, the recombinant FFV Gag protein that was used as substrate contains a thioredoxin region at the amino terminus and starts at residue 154, extending to the carboxyl-terminal residue 514 of FFV Gag (plus a hexa-His tag). The long arrow points to a protein band of 57 kDa that results from the specific cleavage of the p3 protein close to the carboxyl terminus analogous to PFV p3 (also called p4). This assumption was confirmed by assaying the peptide that covers the pre-

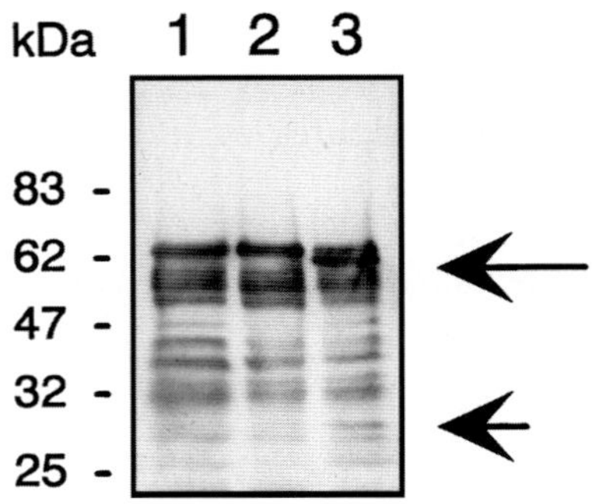

Fig. 3. Western blot analysis of FFV recombinant Gag-thioredoxin of 60 kDa after assay with FFV PR-His. Untreated control, lane *1*; treatment with the inactive FFV PR D/A mutant, lane *2*; and treated with active FFV PR-His, lane *3*. The *long arrow* marks the reaction product of about 57 kDa that corresponds to the carboxyl-terminal Gag cleavage site. The *short arrow* indicates the amino-terminal reaction product generated by a cleavage in the central part of Gag. Antiserum against thioredoxin was used

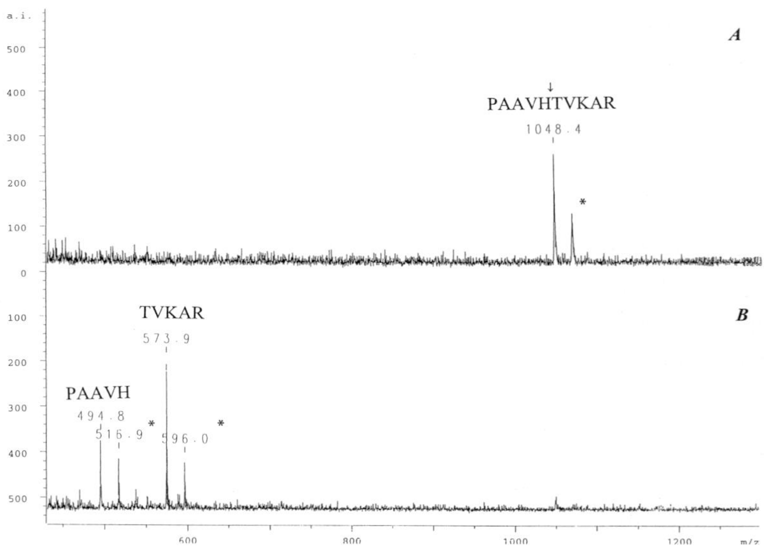

Fig. 4A, B. Identification of the FFV $p48^{Gag}/p3^{Gag}$ cleavage site by MALDI mass spectrometry of the peptide PAAVHTVKAR. **A** The mass peaks of the intact peptide after treatment with the inactive FFV PR D/A mutant. The specific reaction products after treatment with the active FFV PR enzyme are shown in **B**. *Asterisks* mark the Na^+-adducts of both cleavage products

sumed cleavage site at the carboxyl terminus of FFV Gag. The result of combined proteolytic assay and MALDI mass spectrometry is shown in Fig. 4. Recombinant active FFV PR specifically cuts the His-Thr bond located at residues 511/512 of wild-type FFV Gag, whereas the inactive FFV PR D/A mutant did not cleave the decamer peptide PAAVHTVKAR that mimics the native FFV p48/p3^{Gag} site (Fig. 4A). The resulting reaction products precisely correspond in mass to the amino-terminal and carboxyl-terminal peptides and their Na^+ adducts, proving the specific nature of the proteolytic cleavage.

The second specific cleavage product of the terminally truncated FFV Gag protein was detectable at about 26 kDa at a lower concentration (short arrow in Fig. 3). Both controls did not show these two FFV PR products. The second cleavage site should be located in the central part of FFV Gag; the precise site remains to be determined, but an educated guess can be made by analogy with the experimentally determined PFV Gag cleavage sites (Table 1). The assumption is that the second band derives from a specific cleavage in the central part of Gag. The specificity features of FV PRs are discussed below.

Several groups independently reported that PFV Gag proteins ranging in size from 26 to 39 kDa were detectable in infected cells and in purified virions as documented by immunoblot analyses (Netzer et al. 1990; Bartholomä et al. 1992; Netzer et al. 1993; Hahn et al. 1994; Löchelt and Flügel 1996; Luukkonen at al. 1995; Giron et al. 1997; Morozov et al.1997; Pfrepper et al. 1999; Tobaly-Tapiero et al. 1990). In virtually all cases, proteolytic processing was delayed and incomplete as shown by the presence of partially or completely unprocessed pr71^{Gag} and p68^{Gag} proteins. The proteolytic processing of the PFV Gag precursor shown in Fig. 5 is in agreement with the published data.

When comparing individual PFV cleavage sites with each other, the broad range of cleavage efficiency of an individual site is remarkable. The efficiency of cutting a distinct site runs parallel for recombinant proteins and for the corresponding synthetic peptide that covers the cleavage site junction. Whereas the PFV p68^{Gag}/p3^{Gag} and the cleavage sites between RNase H and IN were efficiently cleaved, most other sites were cut at an intermediate or even a lower rate (Pfrepper et al. 1998; Pfrepper 1999; Pfrepper et al. 1999). It is noteworthy that the most efficiently cleaved sites both lie at the carboxyl termini of both precursors. The least efficiently cleaved sites were those located in the central part of PFV Gag. For this reason, we distinguish between primary, secondary,

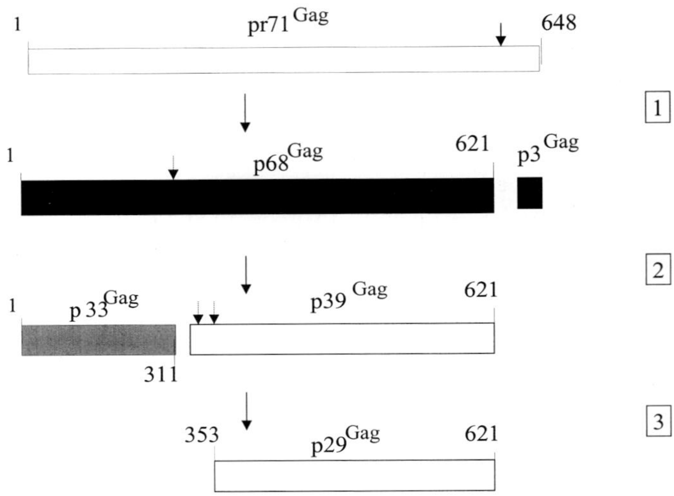

Fig. 5. Schematic diagram of proteolytic processing of PFV Gag precursor. *Solid rectangles* symbolize the different processed forms; *vertical solid arrows* mark the classes of processing events numbered "*1*" as primary, "*2*" as secondary, and "*3*" as tertiary cleavages. *Broken arrows* indicate locations of suboptimal cleavage sites. *Black boxes* show the main FV Gag proteins (for details, see text)

and tertiary cleavage sites of FV proteins. Primary or optimal cleavages and their corresponding reaction products have been observed in different cell types and under normal assay and standard detection conditions. They are cut with efficiencies ranging from about 50% to nearly 100% (Pfrepper et al. 1999). In contrast, secondary and tertiary cleavage events are suboptimal and only observed when sensitive methods of detection are applied and are not found in every cell type analyzed. Whether this distinction has biological significance remains to be seen, but it is useful and illustrates an essential difference in the proteolytic processing of FV precursor proteins compared with those of most RVs. The three classes of cleavage sites shown in Fig. 5 likely reflect the spatial and temporal cleavage events of FV proteolytic processing and are discussed one by one.

Both primary cleavages take place in vitro and in vivo are catalyzed by any one of the three Pro-Pol proteins mentioned above. These cleavages result in the formation of p68^{Gag} plus p3^{Gag} (and p87^{Pol} plus p42^{IN}; see below) . Mutation of the p68/p3 site results in a noninfectious virus

(Enssle et al. 1997; Zemba et al. 1998). The corresponding reaction products were detected in virions and in infected cells. The p3^{Gag} protein has not been detectable so far because of its small size and/or instability.

However, suboptimal secondary and tertiary cleavages were detected by means of PR assays of PFV recombinant Gag proteins and the corresponding peptides that represent the authentic cleavage site junctions. The sites are located in the central part of Gag (Table 1) and result in unusual pattern of Gag processing as two classes of processed forms of Gag proteins are produced, namely small peptides of about 3.5 and 1.5 kDa that are difficult to detect under standard conditions of gel electrophoresis. The second class of processed Gag proteins tends to comigrate between 29 and 34 kDa, because they have similar calculated sizes of about p32 to p33 (Pfrepper et al. 1999). The hallmark of these cleavage events is that they are very inefficient, result in incomplete processing, and are delayed. This holds for recombinant proteins and for peptides that seem to reflect the in vivo situation, where they become detectable by radioactive labeling and after longer exposure of the autoradiograms.

Each of the internal PFV Gag cleavage sites was individually mutated by site-directed mutagenesis of the corresponding P2 sites in the proviral context to confirm the biological significance of these three sites. The P2 positions were chosen because this position is occupied by a conserved bulky and hydrophobic residue, namely either Val or Ile (Table 1). To replace the P2 position by a residue that renders the scissile bond resistant to proteolysis by recombinant PFV PR, Glu was introduced into this position because it had been found that the Glu residue at the carboxyl-terminal region of the PFV-1 PR domain is responsible for the noncleavability of the 143/144 site (Pfrepper et al. 2000, see below). On immunoblot analysis of the Gag mutant Ile310Glu, it was found that virions from the supernatants of baby hamster kidney (BHK-21) cells did not contain a detectable p33^{Gag} protein in contrast to those derived from supernatants of wild-type pHSRV13-transfected BHK-21 cells. (Pfrepper et al. 1999). Similar results were obtained after Western blot analysis of the HFV Gag mutants pVal338Glu and pIle351Glu. On titration of extracellular virus particles on FAB cells (Yu and Linial 1993), the three proviral Gag mutants did not show any infectivity. Together, the data indicate the biological significance of the specificity of the proteolytic processing events of the three internal Gag cleavage sites, with a prominent role of the first site at position 311/312 (Pfrepper et al. 1999). It is assumed that mature FV Gag proteins are required for late steps in maturation and disassembly.

4.2
Pol Processing

To date, only two specific cleavage sites have been reported to occur in the pr125^{Pol} protein. Proteolytic processing of the Pro-Pol-RNAse H and the integrase is a primary cleavage site (Pfrepper et al. 1998), and there is indirect evidence that an intact and active HFV integrase is crucial for virus infectivity (Enssle et al. 1999, Meiering et al. 2000). The cleavage between the Pro-RT and the RNAse H domains occurs with intermediate efficiency but is detectable under standard conditions (Pfrepper et al. 1998). Thus there are three mature HFV Pol proteins, p42^{IN}, p87$^{Pro\text{-}RT\text{-}RH}$, and p65$^{Pro\text{-}RT}$, detectable in virus-infected cells (Pfrepper 1998). A small-size RNAse H band was not detectable, which might be because of rapid degradation.

The specific cleavages in Pol proteins have been recently confirmed by Fenyövalvi et al. (1999), who used a fusion protein of the *E. coli* maltose binding protein and PFV PR domain and synthetic peptides as substrates. In addition, there is evidence for a gradient of increasing enzymatic activities with decreasing length of the native Pro-Pol proteins both in RT and PR activities (Kögel et al. 1995; Pfrepper 1999). Together, these findings indicate that the proteolytic processing at the RNAse H-IN cleavage site is an essential event in FV replication (Baldwin and Linial 1999).

Considering both Pol and Gag processing, secondary cleavages seem to occur at two positions with intermediate efficiency, one at PFV Pol between the RT and RNAse H domain and the other at Gag position 311/312. Comparison of the cleavage site junctions of both Gag and Pol secondary cleavage sites shows that the flanking residues are highly conserved but distinct and partially different from those of the primary sites. An RNAse H protein of low molecular size was not detectable in PFV particles, although a FV RNAse H activity was found in virus-infected cells (Kögel et al. 1995). Tertiary cleavage sites were not found in Pol proteins and Pol-derived peptides.

Why would FV Pro-Pol proteins retain the PR domain at the amino terminus in contrast to other RVs? One reason might be that the retention causes proteolytic processing of FVs to be delayed. Thus the secondary cleavages may take place early in infection and may be essential for disassembly. It has to be considered that the situation of the four enzymatic activities of FV Pol has to be controlled and regulated in view of the fact that Pro-Pol transcripts were detected in relatively high abundance when compared to Gag (Bodem et al. 1996; Holzschu et al. 1998). Although this can easily be explained by the existence of *pol*-specific mRNA in FVs that avoids frameshifting, one must remember that the specific activities of the FV integrase (Pahl and Flügel 1993), reverse transcriptase (Kögel et al. 1995), and protease (Pfrepper 1999) were relatively lower than the corresponding enzymes of HIV-1. A lower level of specific PFV PR activity would circumvent the problem of premature proteolytic processing.

It is well known that one crucial factor is the critical threshold concentration of RV PRs required to be active (Vogt 1997). This is due to the fact that the PR, RT, and IN domains must dimerize as mentioned above. In this context, it is of great relevance to focus on the role of the hinge region between the FV PR and the RT domain, because a PFV PR protein does not seem to occur in a processed form of small molecular size. This points to the issue of the dimeric nature of the active PFV RT. As two native and mature forms of Pro-Pol exist, namely $p87^{Pro\text{-}RT\text{-}RH}$ and $p65^{Pro\text{-}RT}$, this suggests that two different homodimers and one heterodimer of active RT may be formed besides the immature $pr125^{Pol}$. It remains to be analyzed which form is the predominant and active species for reverse transcription in vivo. In addition, there is no precedent of a connecting region between these two domains in other RV Propol proteins with respect to both size and structure of this subdomain, except for that of MLV (Georgiadis et al. 1995).

4.3
Comparative Analysis of Wild-Type FFV and PFV Proteases Mutated Close to the Catalytic Site Residues in Proviral Context

Because all known RV PRs contain an apparently invariant Gly residue in the active center triad D-S/T-G, and because the FFV PR carries a Gln at this position, we assessed whether or not this change is required for proteolytic activity. To test any phenotypic changes, we mutated the Q23 in the FFV PR domain to Gly, and, in addition, the corresponding Gly of the PFV PR domain to Gln in the context of the corresponding proviral DNA.

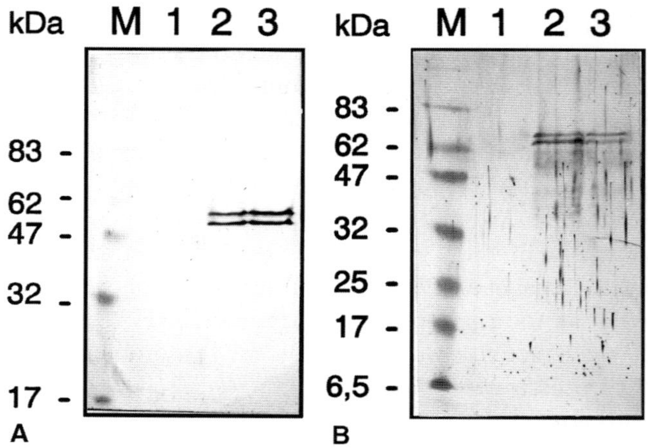

Fig. 6A, B. Comparative analysis of the phenotype of Gag processing by wild-type and mutant FFV and PFV PRs in the proviral context. **A** Western blot analysis of wild-type FFV Gag protein with the FFV PR Q23G mutant in the context of infectious pFeFV7 proviral DNA. CRFK cells were transfected with pUC18 as control, lane *1*; pFeFV7, lane *2*; and pFeFV7 mutant G23Q, lane *3*. Cell extracts were reacted with antiserum against the amino-terminal part of FFV Gag. **B** Western blot analysis of wild-type PFV Gag protein with the PFV PR Q26G mutant in the context of infectious pHSRV13 proviral DNA. BHK-21 cells were transfected with pUC18 as control, lane *1*; pHSRV, lane *2*; and pHSRV13 mutant G26Q, lane *3*. Cell lysates were reacted with an antiserum against the central part of PFV Gag

After cloning by combining fusion PCR and subsequent three-component ligation, the resulting mutants in the proviral context were characterized (Pfrepper 1999). Nucleotide sequencing confirmed the mutation introduced close to the catalytic site. Transfection of the pFFV7 mutant Q23G into Crandell feline kidney (CRFK) cells showed a surprisingly small influence on FFV Gag processing as shown in Fig. 6A. Overall, a slight decrease of proteolytic processing of the mutant versus wild-type infectious clones was detectable both for Gag and for Pol processing (Pfrepper 1999).

When the activity of the recombinant and purified FFV His-tagged mutant PR was assayed with the decamer peptide that covers the p48/p3^{Gag} junction (see above), again a corresponding decrease to about 72% compared to the wild-type FFV PR was found (Pfrepper 1999). This result is in agreement with the data described above for the infectious FFV provirus.

The converse experiment, in which the Gly26 of the PFV PR domain was changed into a Gln residue, was also carried out with a similar approach for cloning. The resulting infectious pHSRV13-derived clone was transfected into BHK-21 cells and analyzed by Western blotting. The result shown in Fig. 6B revealed that Gag processing was marginally reduced compared to wild-type pHSRV13 (Pfrepper 1999). The virus titer was found to be at 3×10^4 2 days after transfection (compared to 1×10^5 for wild type).

Three different PFV Gag peptides that cover cleavage site junctions were subjected to purified mutant PFV PR mutant G26Q, and the cleavage site that corresponds to PFV $p68/p3^{Gag}$ was found to be cleaved with an efficiency decreased by about 30% (Pfrepper et al. 1999). This result may explain why the infectious proviral clone had a lower virus titer in the FAB assay.

5
Specificity of Foamy Viral Protease

The experimentally determined and predicted cleavage sites of primate and nonprimate FV PRs reveal a remarkable resemblance at the P2 and P2' positions with either Val and Ile (or Leu or Thr) as the predominant residues (Table 1). This holds true irrespective of whether it is a primary, secondary, or tertiary site. The rule that Val occupies both P2 positions is valid in all cases of $68^{Gag}/p3^{Gag}$ cleavage sites of six different FVs (Table 1). The RNAse H/IN sites also have a Val at the P2 but not at the P2' positions, where Asn is the most frequent amino acid. The P1 positions prefer hydrophilic residues with few exceptions.

Phe and Tyr can occur at these positions in the case of the ProRT/RNAse H cleavage site because it coincides with a Mg^{2+}-binding site of the RNAse H domain. At most positions, i.e., at P3, P4, P3', and P4' a relatively high variability of the flanking residues is observed. In general, the following conclusions can be drawn: (1) the cleavage site junctions are FV-specific and quite different from the corresponding sites of other RV proteins; (2) they surprisingly resemble the cleavage sites of the yeast retrotransposons Ty1 and Ty3 PRs, as evidenced by the hydrophobicity profiles (Merkulov et al. 1996). The P1 and P1' positions have a marked preference for hydrophilic residues in contrast to those of most RV PRs. Despite this apparent similarity in substrate specificity between the Ty PRs and the PFV PR, four out of six Ty peptide substrates were reported to be resistant to cleavage by the PFV PR (Fenyövalvi et al. 1999). The data indicate that other specificity determinants within the PR binding pocket have to play a role. The exceptionally high conservation of the P1' to P4' positions of the RTRNAse H cleavage site is due to the fact that these four residues coincide with one of the Mg^{2+} binding sites required for RNase H activity (Table 1).

Because peptide substrates that cover the cleavage site junctions have been useful to analyze the influence of individual residues, we have analyzed the corresponding peptide of the carboxyl terminus of the PFV PR domain and found it to be completely resistant against proteolysis by PFV (Pfrepper et al. 2000). In contrast, when the Glu142 residue at the P2 position was mutated to Val, the mutated peptide was efficiently cleaved (Pfrepper et al. 2000). Naturally occurring and mutant forms of the cleavage-resistant peptide were therefore analyzed by circular dichroism to determine whether differences existed in the secondary structures of the peptides that did or did not serve as substrates. The data show that differences in the secondary structure of the native and mutant peptides analyzed do not seem to play a crucial role for cleavage site specificity in PFV PR and might be controlled by the formation with the viral enzyme (Pfrepper et al. 2000).

Site-directed mutagenesis of E54 to V54 within the presumed flap site of the PFV PR clearly showed that this residue is a specificity determinant for the PFV PR enzyme (Pfrepper et al. 2000). The mutant PFV PR enzyme was purified, and the peptide PRAVNTVTQR was used as substrate that covers the primary cleavage site of $pr71^{Gag}$. Analysis of the reaction product showed that it was completely resistant to cleavage by the PFV PR E54V mutant in contrast to cleavability by the wild-type enzyme (Pfrepper et al. 2000). Other peptides that present PFV cleavage site junctions were nonspecifically cleaved with the PFV PR mutant E54V PR. These data clearly indicate that residues that line the substrate binding site of the FV PRs are a crucial part of the factors that contribute to the recognition of an individual specific cleavage site. The requirements for an individual peptide to be susceptible to specific proteolysis by FV PRs can be summarized as follows: (1) A bulky and hydrophobic residues has to occupy both P2 positions. (2) The junction and the adjacent sequences have to be located in an extended β-sheet that provides for accessibility to the enzyme. (3) There should be no charge repulsion between residues of cleavage site junction and the side chains of the flap or binding pocket of the FV PR.

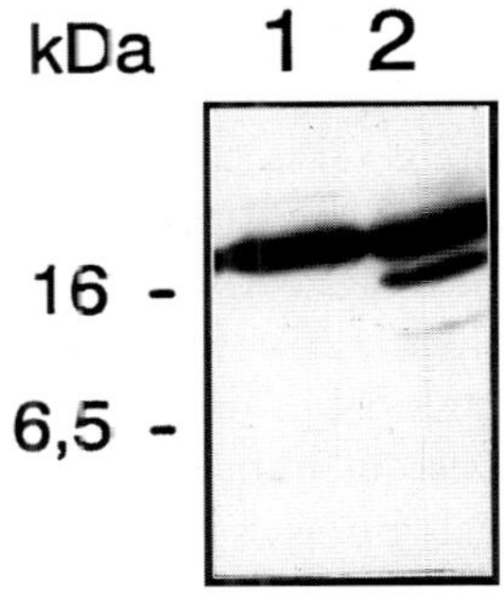

Fig. 7. Autocatalytic degradation of recombinant PFV PR-His. Western blot analysis of purified HFV PR at 37°C. Antiserum against the Pro-Pol domain was used as described previously (Pfrepper et al. 1998). Defective HIS-PR mutant (D/A) as control, lane *1*; active HIS-PR, lane *2*

6
Regulation of Foamy Viral Proteases

Several reports provide evidence for the influence of viral factors on the control of proteolytic activity of different RV PRs (Vogt 1997). For HIV-1 and equine infectious anemia virus (EIAV) PRs it has been shown that the specificity of distinct cleavage sites in the nucleocapsid domain of Gag depends on the binding of viral mRNA (Sheng et al. 1997). In addition, hydrophilic peptides from the transframe region of HIV-1 have been reported to block the HIV-1 PR (Louis et al. 1998; Paulus et al. 1999). By deleting distinct PFV U5 sequences, Henkelein et al. (2000) recently reported a negative effect on the proteolytic processing of Gag by Pol. However, native proteins that inhibit the activity of FV PR activity remain to be identified.

There seems to be yet another level of regulating retroviral PR activity. This is illustrated by feline immunodeficiency virus (FIV) PR, for which it has been reported that autocatalysis can result in the removal the first six amino-terminal residues (Laco et al. 1997). This substantially contributes to a higher stability of the enzyme. Other reports concern the autocatalytic degradation of the HIV-1 PR that leads to complete loss of enzymatic activity (Rose et al. 1993; Louis et al. 1999). To explore whether purified PFV PR-His protein cleaves itself, the enzyme was incubated for 16 h and the reaction products were separated by gel electrophoresis and visualized by enhanced chemiluminescence. Besides the 16-kDa starting material, three additional bands of 13 (intense), 10 (weak), and 6 kDa (very weak) were detected (Fig. 7) (Pfrepper et al. 1998). The intense band of 13 kDa was subjected to amino-terminal sequencing and found to start at (NH_2)-YVTFK, which corresponds to a cleavage site located at the central Tyr-Tyr 60 and 61 of the PFV PR domain. This central cleavage resulted in two reaction products of 60 and 87 residues (6 and 10 kDa, respectively). The second cleavage occurred near the carboxyl terminus, yielding the 13-kDa band and a 3-kDa peptide. Amino-terminal sequencing of the 13-kDa band resulted in the amino-terminal residues of the PFV PR domain, indicating that the cleavage must have occurred close to the its carboxyl terminus. The *m/z* value of the peak determined by MALDI-mass spectrometry was 2843.3, which is in agreement with the peptide sequence FVKYDNLWQHWENLE-$(H)_6$ at the carboxyl terminus of PFV PR-His. Thus the scissile bond is at Leu130 and F131. Whereas the specific cleavage at the central cleavage site (Tyr60-Tyr61) led to a complete loss of enzymatic activity, the carboxyl-terminal processing did not. Whether these events have physiological relevance remains to be seen.

7 Summary and Conclusions

The viral protease is an enzyme with unprecedented structural and functional features that is required for retroviral replication. The remarkable property of FV PRs that are fully active as long PR-Pol proteins is a novel feature of retroviral proteases. The occurrence of a Gln residue instead of Gly close to the catalytic center D-S/T-G of the enzymatically active FFV PR is unique. The level and timing of regulation of Gag and Pol processing remain to be analyzed.

The specific $p125^{Pol}$, $p87^{Pol}$, and $p65^{Pol}$ forms of active reverse transcriptase seem to reside in virions and to be active as proteases. Hence there are multiple forms of active FV PRs, an intriguing feature of all RVs that might contribute to cleavage site specificity. The conventional patterns of retroviral MA, CA ,and NC Gag proteins are not observed in FV. One reason is that these Gag domains do not occur in FVs, because the Gag hallmarks, e.g., the major homology region and the typical zinc finger-like sequences of the retroviral NCs are absent. Gag processing is incomplete and delayed, suggesting that it may occur after cell entry. The internal cleavage sites, once mutated in the proviral context, are not infectious. The existence of high-molecular-mass Pro-Pol proteins of relative low specific activity likely contributes to the very low level of proteolytic processing at internal sites of Gag. The influence of viral and cellular factors on the regulation of FV PR activity is unknown. It will be important to analyze details of the special mechanisms and features of Gag maturation and assembly.

Two recent reports indicate novel aspects of regulation of RV PRs. Both reports indicate that activation of the PRs of HIV-1 and type D RVs within immature capsids depend on the redox state of distinct Met and Cys residues in the PR protein sequences and occur in the absence of viral budding (Davis et al. 1999; Parker and Hunter 2001).

References

Babe LM, Craik CS (1997) Viral proteases: evolution of diverse structural motifs to optimize function. Cell 91:427–430

Baldwin DN, Linial ML (1999) Proteolytic activity, the carboxy-terminus of Gag, and the primer binding site are not required for Pol incorporation into foamy virus particles. J Virol 73:6387–93

Bartholomä A, Muranyi W, Flügel RM (1992) Bacterial expression of the capsid antigen domain and identification of native Gag proteins from spumavirus-infected cells. Virus Res 23:27–38

Bernstein NK, Cherney MM, Loetscher H, Ridley RG James MNG (1999) Crystal structure of the novel aspartic proteinase zymogen proplasmin II from *Plasmodium falciparum*. Nature Struct Biol 6:32–37

Bodem J, Löchelt M, Winkler I, Flower RT, Delius H, Flügel RM (1996) Characterization of the spliced pol transcript of feline foamy virus: the splice acceptor of the pol transcript is located in gag of foamy viruses. J Virol 70:9024–9027.

Bodem J, Löchelt M, Flügel RM (1998) Detection of subgenomic cDNAs and mapping of feline foamy virus mRNAs reveals complex patterns of transcription. Virology 244:417–426

Coffin JM, Hughes SH, Varmus HE (eds) (1997) Synthesis, assembly and processing of viral proteins. Cold Spring Harbor Laboratory Press, New York, pp 263–334

Davis et al. (1999) J Virol 73:1156

Dunn B (1997) Splitting image. Nat Struct Biol 4:969–972

Enssle J., Jordan I, Mauer B, Rethwilm A. (1996) Foamy virus reverse transcriptase is expressed independently from the Gag protein. Proc Nat. Acad Sci USA 93:4137–4141

Enssle J., Fischer N, Moebes A, Mauer B, Smola U, Rethwilm A. (1997) Carboxy-terminal cleavage of the human foamy virus gag precursor molecule is an essential step in the viral life cycle. J Virol 71:7312–7317

Enssle J., Moebes A, Henkelein, M., Panhuysen, M., Mauer, B., Schweizer, M., Neumann-Haefelin, D., Rethwilm A. (1999) An active human foamy virus integrase is required for viral replication. J Gen Virol 80: 1445–1452

Fenyövalvi G, Bagossi P, Copeland TD, Orozlan S, Boross P, Tözser J (1999) Expression and characterization of the human foamy virus proteinase. FEBS Lett 462:397–401

Georgiadis MM, Jessen SM, Ogata CM, Telesnitzky A, Goff SP, Hendrickson WA (1995) Mechanistic implications from the structure of a catalytic fragment of Moloney murine leukemia reverse transcriptase. Curr Biol 3:879–892

Hahn H, Baunach G, Bräutigam S, Mergia A, Neumann-Haefelin D, Daniel MD, McClure MO, Retwilm A (1994) Reactivity of primate sera to foamy virus Gag and Bet proteins. J Gen Virol 75:2635–26344

Heinkelein M, Thurow J, Dressler M, Imrich H, Neumann-Haefelin D, McClure MO, Rethwilm A (2000) Complex effects of deletion in the 5' untranslated region of primate foamy virus on viral gene expression and RNA packaging. J Virol 74:3141–3148

Holzschu DL, Delaney MA, Renshaw RW (1998) The nucleotide sequence and *pol* mRNA levels of the nonprimate spumavirus bovine foamy virus. J Virol 72:2177–2182

Khan AR, James MNG (1998) Molecular mechanisms for the conversion of zymogens to active proteolytic enzymes. Protein Sci 7:815–836

Kögel D, Aboud M, Flügel RM (1995) Molecular biological characterization of the human foamy virus reverse transcriptase and ribonuclease H domains. Virology 213:97–108

Konvalinka J, Löchelt M, Zentgraf H, Flügel RM, Kräusslich H-G (1995) Active spumavirus proteinase is essential for virus infectivity but not for formation of the Pol polyprotein. J Virol 69:7264–7268

Kotler M, Katz RA, Danho W, Leis J, Skalka AM (1988) Synthetic peptides as substrates and inhibitors of a retroviral protease. Proc Natl Acad Sci USA 85:4185–4189

Laco GS, Fitzgerald MC, Morris GM, Olson AJ, Kent SBH, Elder JH (1997) Molecular analysis of the feline immunodeficiency virus protease: generation of a novel form of the protease by autoproteolysis and construction of cleavage-resistant proteases. J Virol 71:5505–5511

Lecellier CH, Saïb A (2000) Foamy viruses: between retroviruses and pararetroviruses. Virology 271:1–8

Lee AY, Gulnik SV, Erickson JW (1998) Conformational switching in an aspartatic proteinase. Nat Struct Biol 5:866–871

Linial ML (1999) Foamy viruses are unconventional retroviruses. J Virol 73:1747–1755

Löchelt M, and Flügel RM (1995) The molecular biology of primate spumaviruses. In: Levy JA (ed) The Retroviridae, Vol 4, pp 239–292, Plenum Press, New York

Löchelt M, Flügel RM (1996) The human foamy virus *pol* gene is expressed as a Pro-Pol polyprotein and not as a Gag-Pol fusion protein. J Virol 70:1033–1040

Lois JM, Dyda F, Nashed NT, Kimmel AR, Davies DR (1998) Hydrophilic peptides from the transframe region of Gag-Pol inhibit the HIV-1 protease. Biochemistry 37:2105–2110

Louis JM, Clore GM, Gronenborn AM (1999) Autoprocessing of HIV-1 protease is tightly coupled to protein folding. Nature Struct Biol 6:868–875

Luukkonen BGM, Tan W, Fenyö EM, Schwartz S (1995) Analysis of cross reactivity of retrovirus proteases using a vaccinia virus-T7 RNA polymerase-based expression system. J Gen Virol 76:2169–2180

Meiering CD, Comstock KE, Linial Ml (2000) Multiple integrations of human foamy virus in persistently infected human erythroleukemia cells. J Virol 74:1718–1726

Merkulov GV, Swiderek KM, Brachmann B, Boeke JD (1996) A critical proteolytic cleavage site near the C-terminus of the yeast retrotransposon Ty1 Gag protein. J Virol 70:5548–5556

Morozov VA, Copeland TD, Nagashima K, Gonda MA, Oroszlan S (1997) Protein composition and morphology of human foamy virus intracellular cores and extracellular particles. Virology 228:307–317

Netzer K-O, Rethwilm A, Maurer B, ter Meulen V (1990). Identification of the major immunogenic structural proteins of human foamy virus. J Gen Virol 71:1237–1241

Netzer K-O, Schliephake A, Maurer B, Watanabe R, Aguzzi, A, Rethwilm A (1993) Identification of pol-related gene products of human foamy virus. Virology 192:336–338

Pahl A, Flügel RM (1993) Endonucleolytic cleavages and DNA-joining activities of the integration protein of human foamy virus. J Virol 67:5426–5434

Parker and Hunter (2001) Proc Natl Acad Sci USA 98: 14361

Paulus C, Hellebrand S, Tessmer U, Wolf H, Kräusslich H-G, Wagner R (1999) Competitive inhibition of HIV-1 protease by the Gag-Pol transframe protein. J Biol Chem 274:21539–21543

Pfrepper K-I, Rackwitz H-R, Schnölzer M, Heid H, Löchelt M, Flügel RM (1998) Molecular characterization of proteolytic processing of the Pol proteins of human foamy virus reveals novel features of the viral protease. J Virol 72:7648–7652

Pfrepper K-I, Rackwitz H-R, Schnölzer M, Heid H, Löchelt M, Flügel, RM (1997) Expression and molecular characterization of an enzymatically active recombinant human spumaretrovirus protease. Biochem Biophys Res Commun 237:548–553

Pfrepper K-I, Rackwitz H-R, Schnölzer M, Heid H, Löchelt M, Flügel RM (1999) Molecular characterization of proteolytic processing of the Gag proteins of human spumavirus J Virol 73:7907–7911

Pfrepper K-I (1999) Molecular characterization of the viral protease and proteolytic processing of the Gag and Pol proteins of the human spumaretrovirus. PhD thesis, Faculty of Biology, University of Heidelberg

Pfrepper K-I, Reed J, Rackwitz H-R, Schnölzer M, Flügel RM (2000) Characterization of peptide substrates and viral enzyme that affect the cleavage site specificity of the human spumaretrovirus proteinase. Virus Genes 22:61–72

Rose JR, Salto R, Craik CS (1993) Regulation of autoproteolysis of the HIV-1 and HIV-2 proteases with engineered amino acid substitutions. J Biol Chem 268:11393–11945

Rost B, Sander C (1994) Combining evolutionary information and neural networks to predict protein secondary structure. Proteins 9:55–72

Sheng N, Petit SC, Tritch RJ, Ozturk DH, Rayner MM, Swanstrom R, Erickson-Virtanen S (1997) Determinants of the human immunodeficiency virus type 1 p15NC-RNA interaction that affect enhanced cleavage by the viral protease. J Virol 71:5723–5732

Tobaly-Tapiero J, Santillana-Hayat M, Giron M-L, Guillemin MC, Rozain F, Peries J, Emanuel-Ravier R (1990) Molecular differences between two immunologically related spumaretroviruses: the human prototype HSRV and the chimpanzee isolate SFV6. AIDS Res Human Retrovir 6:951–957

Tobaly-Tapiero J, Bittoun P, Neves M, Guillemin MC, Lecellier C-H, Puvion-Dutilleul F, Gicquel B, Zientara S, Giron M-L, de The H, Saïb A (2000) Isolation and characterization of an equine foamy virus. J Virol 74:4064–4077

Vogt and Eisenman (1973) Interaction of a large polypeptide precursor of avian oncornavirus Gag protein. Proc Natl Acad Sci USA 70:1734–1738

Vogt VM (1997) Proteolytic processing and particle maturation. Curr Topics Microbiol Immunol 214:95–131

Winkler I, Bodem J, Haas L, Zemba M, Flower RT, Delius H, Flügel RM, Löchelt M(1997). Characterization of the genome of feline foamy virus and its proteins shows distinct features different from primate spumaviruses. J Virol 71:6727–6741

Wlodawer A, Erickson J (1993) Structure-based inhibitors of HIV-1. Annu Rev Biochem 62:543–585

Wlodawer A, Gustchina A, Reshetnikova L, Lubkowski J, Zdanov A, Hui YH, Angleton L, Farmerie WG, Goodenow MM, Bhatt D, Zhang L, Dunn BM (1995) Structure of an inhibitor complex of the proteinase from feline immunodeficiency virus. Nat Struct Biol 2:480–488

Wlodawer A, Gustchina A (2000) Structural and biochemical studies of retroviral proteases. Biochim Biophys Acta 1477:16–34

Wolfe MS, Xia W, Ostazewsli BL, Diehl, Kimberly WT, Selkoe DJ (1999) Two transmembrane aspartates in presenilin-1 required for presenilin endoproteolysis and γ-secretase activity. Nature 398:513–517

Yan R, Bienkowski MJ, Shuck ME, Miao, H, Tory MC, Pauley AM, Brashler JR, Stratman NC, Matthew WR, Buhl AE, Carter DB, Tomaselli AG, Parodi LA, Heinrikson RL, Gurney ME (1999) Membrane-anchored aspartyl protease with Alzheimer's disease γ-secretase activity. Nature 402,:531–537

Yu SF, Baldwin DN, Gwynn SR, Yendapalli S, Linial ML (1996) Human foamy virus replication—a pathway distinct from that of retroviruses and hepadnaviruses. Science 271:1579–1582

Yu, S.F., Linial, M., L. (1993) Analysis of the role of the bel and bet open reading frames of the human foamy virus by using a new quantitative assay. J Virol 67:6618–6624

Zemba M, Wilk T, Rutten T, Wagner A, Flügel RM, Löchelt M (1998). The carboxy-terminal $p3^{Gag}$ domain of the human foamy virus Gag precursor is required for efficient virus infectivity. Virology 247:7–13

Chapter 4

Particle Assembly and Genome Packaging

M. L. Linial, S. W. Eastman

Abstract Foamy virus (FV) replication is distinct from that of all other retroviruses in many respects, including viral assembly. In fact, the viral assembly pathway is rather similar to that of hepadnaviruses such as hepatitis B virus. Foamy virus Gag does not contain landmark retroviral assembly domains such as the major homology region, Cys-His boxes, or a defined M domain. Like hepadnaviruses, the FV Gag protein is not cleaved and contains arginine-rich regions at the carboxyl terminus. In addition, egress of FV particles requires presence of the envelope glycoproteins. Finally, the *cis*-acting sequences in the FV genome required for

M. L. Linial (✉), S. W. Eastman
Division of Basic Sciences A3–015, Fred Hutchinson Cancer Research Center, 1100 Fairview Avenue N., Seattle, WA 98109, USA
E-mail: mlinial@fhcrc.org

genome incorporation, although poorly defined, differ in location from other retroviruses.

1
Assembly of Viral Particles Through Gag-Targeting and Gag–Gag Interactions

Retroviral assembly is of two general types, which have been called C-type and B/D-type. In the first, typical of gammaretroviruses such as murine leukemia virus (MLV) or lentiviruses such as human immunodeficiency virus (HIV), particles assemble at the cytoplasmic face of the cell plasma membrane. In the second type, used by betaretroviruses such as Mason-Pfizer monkey virus (MPMV), particles assemble at an intracellular location and completed particles are transported to a cell membrane for glycoprotein incorporation and budding. Foamy virus (FV) assembly more closely follows the second intracellular pathway. However, unlike most retroviruses, the budding of capsids requires expression of the Env glycoprotein. In addition, the FVs have an immature

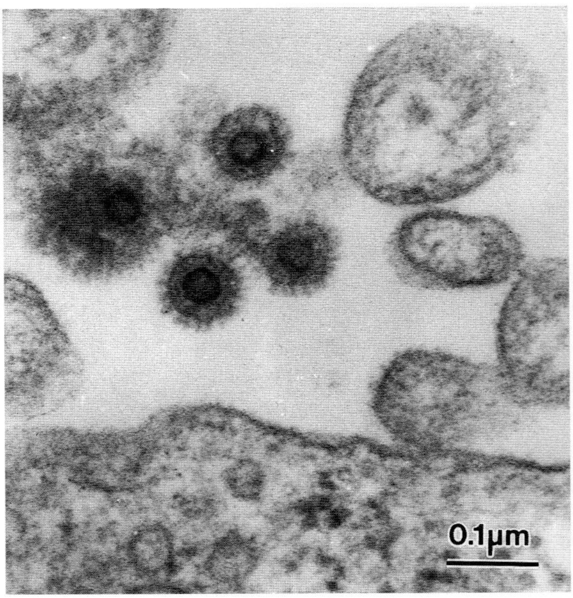

Fig. 1. Electron micrograph of extracellular PFV particles

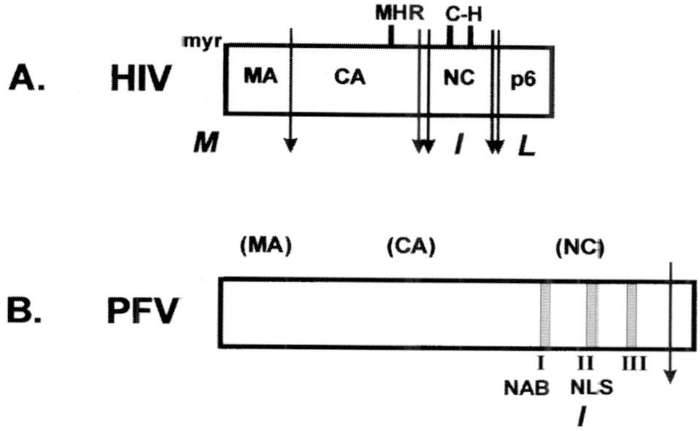

Fig. 2A, B. Gag proteins. The *arrows* indicate the major proteolytic cleavage sites. **A** Conventional retroviral Gag; details shown for HIV. *Black boxes* show the location two Cys-His boxes (*C-H*) and the major homology region, MHR; *myr*, myristylated glycine residue; *MA*, matrix protein; *CA*, capsid protein; *NC*, nucleocapsid protein; *p6*, HIV specific virion protein. *Italicized letters* indicate assembly domains: *M*, membrane domain; *I*, interaction domain; *L*, late domain. **B** The prototypic foamy virus (PFV) Gag. The retroviral domains are shown in *parentheses*. The *gray boxes* labeled *I*, *II*, and *III* represent the glycine-arginine rich GR boxes. *NAB*, nucleic acid binding; *NLS*, nuclear localization signal

morphology (Fig. 1), consistent with a lack of complete protease-mediated processing of the Gag protein (see the chapter by Flügel and Pfrepper, this volume).

1.1 Conventional Retroviruses

To understand how FVs assemble, it is instructive to examine the pathways known for other retroviruses that are much better characterized. For all conventional retroviruses such as HIV, assembly occurs through domains in the Gag precursor protein (Fig. 2A). For these viruses, the Gag precursor protein is not cleaved by the viral protease (usually encoded within the Pol gene) until assembly and budding are complete. Mature particles are composed of three to five cleaved Gag proteins, which always include matrix (MA), capsid (CA), and nucleocapsid (NC). The domains required for capsid formation and budding have been well defined for several viruses and are called the membrane (M), interacting (I), and late (L) domains (reviewed in Swanstrom and Wills 1997). The

M domain is situated at the amino terminus of the Gag protein. For most retroviruses, this domain includes an essential amino-terminal myristylation signal as well as a cluster of basic residues within the first 80 aa, although the avian retroviruses are not myristylated. The M domain is important for targeting and interaction of Gag with the plasma membrane for the alpha- and gammaretroviruses and lentiviruses. An equivalent domain has not been well characterized for the betaretroviruses, which transport preassembled particles to the membrane. The I domain is important for directing the Gag–Gag interactions necessary to mediate particle assembly and maps to the NC domain at the carboxyl terminus of Gag. The I domain is thought to be required for assembly of particles of the proper density (Bennett et al. 1993) and may be coincident with a high concentration of basic residues (Bowzard et al. 1998), although the role of basic residues is controversial (Cimarelli and Luban 2000). Evidence is accumulating that RNA is an integral part of at least C-type Gag–Gag interactions and particle assembly (Campbell and Vogt 1995), and thus it is conceivable that the I domain is important for Gag–RNA binding. Recent work suggests that the binding of the I domain to RNA allows multimerization at other sites within Gag (Sandefur et al. 2000). The I domain is the one retroviral Gag domain that is believed to be a part of the FV Gag protein (Bowzard et al. 1998). Finally, the L domain is important for separation of the nascent retroviral particle from the cell surface. This is a proline-rich motif found at different locations in different retroviruses. Despite differences in amino acid composition, these motifs can substitute for each other in a positionally independent manner (Garnier et al. 1999; Yuan et al. 2000). In the absence of the L domain particles accumulate at the membrane but are not efficiently released. Recent work describes a number of cellular factors within the multivesicular body sorting system that play a role in mediating the L domain function during particle budding. (Garrus et al. 2001; Martin-Serrano et al. 2001; VerPlank et al. 2001).

In addition, conventional retroviruses have two other conserved domains. The first is the major homology region (MHR) found in CA. In HIV, the MHR appears to be necessary for viral assembly (Mammano et al. 1994). In contrast, however, the MHR of Rous sarcoma virus (RSV), an avian retrovirus, is dispensable for viral assembly (Craven et al. 1995). The second is the Cys-His box ($CX_2CX4HX4C$), present in one or two copies in the NC protein of all conventional retroviruses. Cys-His boxes are involved in RNA encapsidation as well as other functions of

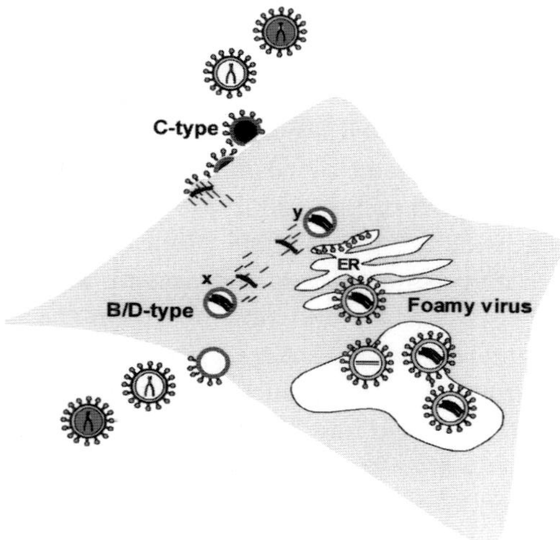

Fig. 3. Retroviral assembly pathways. The rough scheme for assembly of gammaretroviruses (murine leukemia virus), alpharetroviruses (avian leukosis virus), and lentiviruses (human immunodeficiency virus) and betaretroviruses (Mason-Pfizer monkey virus and murine mammary tumor virus) as well as the primate foamy viruses are shown. The *short black lines* indicate nascent Gag proteins and the *long curved lines* viral RNA. The clear intracellular *circles* represent preformed capsids. The extracellular virions with *clear centers* are of the immature form, and mature virions have *gray centers*. The knobs represent glycoproteins. *ER*, endoplasmic reticulum; *x*, assembly site for the B/D type; *y*, assembly site for FV

the life cycle (reviewed in Berkowitz et al. 1996). Both of these domains are absent from FV genomes.

A general scheme for retroviral assembly involves interaction of the Gag precursor protein, through the I domain as well as other regions, with both other Gag molecules and RNA to initiate the assembly process. Assembly occurs at the cellular membrane in the case of alpha- and gammaretroviruses and lentiviruses (C-type) or at a yet unknown intracellular location (x) for betaretroviruses (B/D-type) (Fig. 3). The betaretroviruses contain a dominant cytoplasmic targeting-retention signal (CTRS) near the amino terminus of Gag (Choi et al. 1999). Mutations within this signal redirect assembly to the plasma membrane, a process dependent on myristylation (Rhee and Hunter 1990; Sakalian and Hunter 1999). Conversely, addition of this signal to a C-type virus allows assembly to take place in the cytoplasm (Choi et al. 1999). Thus it appears that conventional retroviruses have the intrinsic capacity to assemble at multiple locations within the cell. In the case of the betaretroviruses, transport of the nascent capsids to the cell membrane occurs through the M domain sequences, which are also responsible for membrane targeting of alpha- and gammaretroviruses and lentiviruses. Myristylation (Myr)-minus Gag proteins of this group are not targeted to the plasma membrane, and capsids fail to assemble (Rein et al. 1986; Spearman et al. 1994), whereas Myr-minus betaretroviruses do assemble but are not transported to the cell surface and remain within the cell (Rhee et al. 1987). Nascent betaretroviruses, gammaretroviruses, and lentivirus particles are immature; the Gag precursor proteins have not yet been cleaved by the viral protease into the individual components. This occurs during or shortly after budding. Maturation of all of these viruses is absolutely required for infectivity.

1.2 Foamy Viruses

Much less is known about FV assembly, and almost all of the available data apply to the primate FVs, in particular the human isolate PFV (also called HFV or SFVcpz[hu]). The alignment of the Gag proteins of the six sequenced FVs is given in Fig. 4. Unlike all other retroviral Gag proteins, FV Gag is not cleaved into MA, CA, and NC components. Proteolytic processing is limited to a single event, resulting in a 3-kDa carboxyl-terminal truncation of the 71-kDa precursor protein. The overall conservation between the isolates is low and is concentrated near the amino termini and within a central region. At the carboxyl termini there are high concentrations of glycine and arginine residues, although the specific amino acid sequences are not well conserved. These regions have been termed GR boxes, analogous to those found in the core protein of the reverse transcriptase (RT)-encoding hepatitis B virus (HBV) (Schliephake et al. 1994; Fig. 2B). GR Box I is important for nucleic acid binding (Yu et al. 1996b), and GR box II contains a nuclear localization signal required to transport nascent Gag proteins to the nucleus (Schliephake et al. 1994). The NLS is not required for growth of virus in tissue culture (Yu et al. 1996b), and the role of Gag nuclear localization in the viral life cycle is unknown. The NLS could play a role in acquisition of additional proviral copies through an intracellular pathway (Heinkelein et al. 2000a; Meiering et al. 2000).

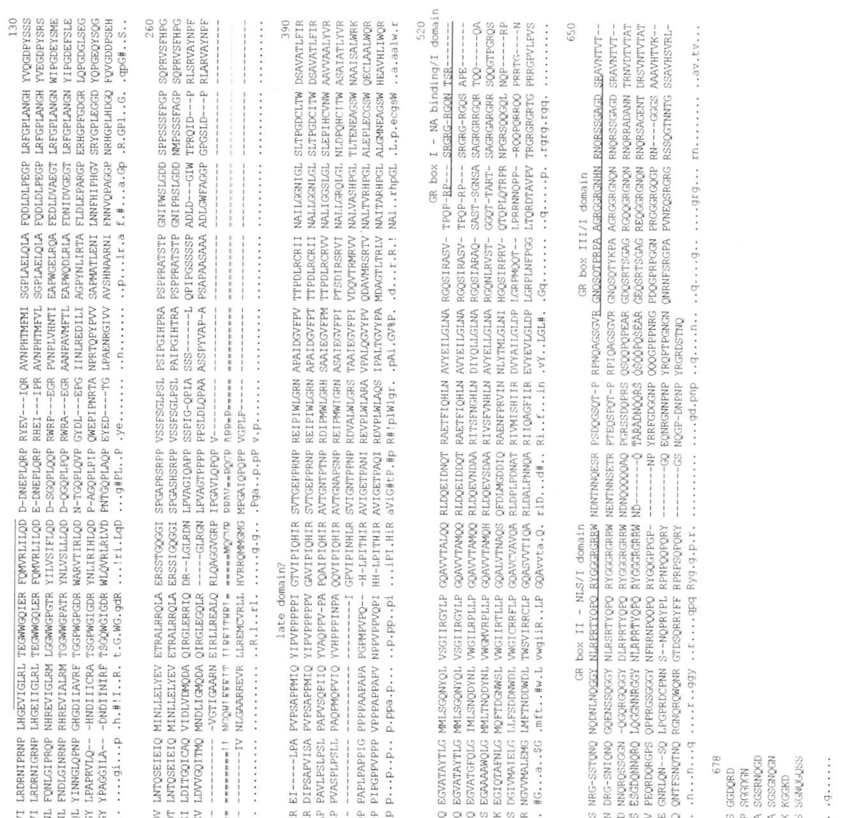

Fig. 4. Alignment of the Gag proteins of seven sequenced foamy viruses. The consensus amino acids (aa) are listed at the bottom of each set with *capital letters* indicating completely conserved aa and the small letter consensus aa. *Dashes* indicate aa missing in the isolate. The putative CTRS is shown by a *line* over the PFV sequence. The three GR boxes are indicated by the *lines* under the PFV sequence. The putative late domain and conserved arginine (*R*) important for morphogenesis are indicated in *bold* in the PFV sequence

The role of other conserved domains in Gag is not known. The FV Gag proteins do not contain a consensus myristylation signal at the amino terminus and are not modified by myristylation (S. Eastman and M. Linial, unpublished results). Although plasma membrane targeting of Gag does not take place, centrosomal targeting of incoming particle-associated Gag has been observed (Saïb et al. 1997). Recent data provide evidence for an interaction between FV Gag and the cytoplasmic light chain of dynein, LC8, a cellular factor usurped by a number of other RNA viruses, including rabies virus, for trafficking viral proteins within the cytoplasm of infected cells (Raux et al. 2002; Petit et al. 2003). This interaction between Gag and LC8 is presumably responsible for targeting capsids to the microtubule organizing complex, thereby allowing the subsequent disassembly and nuclear import of the viral cDNA. A region between amino acids 150 and 250 of Gag is responsible for this centrosomal targeting (Petit et al. 2003).

Interestingly, a region from amino acids 301–476 of PFV Gag, which contains most of the conserved amino acids, can be deleted without abrogating the production of extracellular particles (S. Eastman and M. Linial, unpublished results). Thus this region is unlikely to contain an I domain or any other assembly domains such as the MHR, which appear to be required for HIV-1. In addition, PFV Gag contains a PPPPPI sequence, similar to a consensus L domain sequence at amino acid 359 (Fig. 4), but it is not known whether this sequence is important for budding. The poor release of virus from cells could suggest that any L domain would be part of an inefficient pathway. In some respects, the FV Gag protein is more similar to the core protein of the hepadnaviruses (Table 1). For example, hepadnaviral Gag proteins are not cleaved and contain a large number of carboxyl-terminal arginines.

Infection of fibroblast cells in tissue culture with wild-type FV produces particles that remain mostly cell associated; only about 5% of the infectious particles are found in the culture supernatant (Yu et al. 1993). FVs assemble in the cytoplasm in a manner similar to that of the betaretroviruses. In the case of the primate FV, preformed particles then localize to the endoplasmic reticulum, where they are enveloped (Fig. 3). However, in the absence of Env protein, particles do not bud from the cell (Baldwin et al. 1998; Fischer et al. 1998) but remain completely cell associated, suggesting that a specific interaction between Gag and Env is required for particle egress. Retention of viral Env in the ER requires a dilysine motif in Env, and PFV particles mutant in this domain bud to a

Table 1. Comparison of the properties of the structural proteins between conventional retroviruses, foamy viruses, and hepadnaviruses

Virus group	Retrovirus	Foamy virus	Hepadnavirus
Structural protein	Gag	Gag	Core
Protease cleavage products	MA	Gag	None
	CA	p3	
	NC		
1–2 others			
NA binding domains	NC CH boxes + basic aa	GR box I	R at carboxyl terminus

For abbreviations used, see Fig. 2

greater extent from the plasma membrane (Goepfert et al. 1997, 1999). Equine (EFV) and bovine (BFV) foamy virus Env sequences seem to lack a functional ER retrieval signal (Renshaw and Casey 1994, Tobaly-Tapiero et al. 2000). However, whereas EFV has been reported to bud exclusively from the plasma membrane, earlier analysis of BFV morphogenesis also found budding at intracytoplasmic membranes (Dermott et al. 1971; Tobaly-Tapiero et al. 2000). Thus it appears that the Gag-Env interaction may compensate for the lack of a membrane targeting signal. If a known myristylation sequence is placed at the amino terminus of PFV Gag, viral particles can bud in the absence of Env protein, consistent with this notion (Eastman and Linial 2001). The Env domain required for this interaction has recently been identified and resides at the amino terminus of the Env signal peptide (Lindemann et al. 1997, 2001; see the chapter by Lindemann and Goepfert, this volume). The specific Gag domains are unknown, although cryoelectron microscopy and data from surface plasmon resonance experiments using a bacterially expressed and purified amino-terminal domain (matrix equivalent) of FFV Gag and the amino-terminal region of the Env signal peptide have shown that these proteins specifically interact in vitro in a sequence-dependent manner (Wilk et al. 2001).

The mechanism mediating intracellular capsid assembly for retroviruses is unknown, but protein targeting and host cellular factors are thought to be responsible. The FV genomes contain a consensus CTRS domain (Tobaly-Tapiero et al. 2000). In fact, mutation of a highly conserved arginine central to this sequence (Fig. 4, R in bold) inhibits capsid assembly and prevents particle release (Eastman et al. 2001). The domains of FV Gag involved in capsid assembly are unclear. Recent experiments have described a region in the amino-proximal portion of Gag, aa 131–162, possessing a coiled-coil motif that is possibly involved in mediating capsid assembly (Tobaly-Tapiero et al. 2000). These authors generated fusions of this region with the nuclear protein PML and showed that, upon introduction of Gag proteins lacking an NLS, the heterologous protein was redirected to the cytoplasm, implying that this region of Gag may act as a Gag–Gag interaction domain. Consistent with this hypothesis, introduction of this mutation into the provirus resulted in a complete block to particle release (Tobaly-Tapiero et al. 2000). These data are suggestive of the presence of an assembly domain but require a more thorough analysis to be conclusive.

Together, these results suggest the following FV assembly pathway. A CTRS-like signal at the amino terminus is required to target and concentrate Gag proteins at an intracellular site to initiate Gag–Gag interactions and capsid formation. Assembly into particles exposes a Gag domain that can then interact with the viral Env protein signal peptide located at the ER membrane. This interaction leads to particle transport into the ER or derived vesicles and ultimately to egress from the cell.

2
Pol Synthesis and Incorporation

In most retroviruses, the Pol protein encodes three different enzymatic activities, protease (PR), reverse transcriptase (RT) with its associated RNAseH activity, and integrase (IN). The Pol protein must be packaged into particles to allow PR cleavage of the Gag precursor for maturation, to provide RT to newly infected cells where it makes cDNA copies of genomic RNA for integration, and to deliver IN to integrate the cDNA into the cell chromosome to create the provirus. Activation of any of these enzymatic activities in the host cell before sequestration into particles could be deleterious to the host and ultimately to the success of the virus. Therefore, specific mechanisms to package inactive Pol into particles exist for all RT-encoding viruses. In addition, retroviruses generally synthesize both Gag and Pol from the same genomic length RNA but require many more Gag molecules than Pol molecules per virion.

2.1 Conventional Retroviruses

In the six other genera of retroviruses, Pol proteins are synthesized as Gag-Pol fusion proteins by one of two mechanisms (reviewed in Jacks 1990). The first is via ribosomal frameshifting and is exemplified by HIV. The Pol protein reading frame is −1 relative to Gag, and during Gag translation ribosomes occasionally slip backward one nucleotide to create a Gag-Pol protein. The second mechanism is represented by MLV and occurs via suppression of termination of a stop codon at the end of Gag. Both of these mechanisms are inefficient, leading to a ratio of Gag to Gag-Pol of about 10–20:1. Gag-Pol proteins are targeted to assembly sites by the same mechanisms as Gag proteins, discussed above.

2.2 Foamy Virus Pol Synthesis

FV Pol is in a +1 reading frame relative to Gag, eliminating the possibility of both of the normal retroviral Gag-Pol synthesis mechanisms. There is no evidence for a stable Gag-Pol fusion protein in cells, even in a protease-negative virus (Konvalinka et al. 1995). This leaves the possibility that Pol is synthesized independently of Gag. There are at least three ways this could occur (Fig. 5). (1) Use of an internal ribosomal entry site (IRES) to synthesize Pol from genomic RNA. (2) Use of an internal promoter encoded within Gag to drive expression of a Pol mRNA. This would be similar to the promoter in Env used for transcription of Tas and Bet mRNAs. (3) Use of a Pol-specific spliced mRNA. Strong evidence was found for the presence of a PFV spliced mRNA that is required for synthesis of the Pol protein (Yu et al. 1996a). Three types of experiments were done. First, with RT-PCR and primers specific to the *gag* and *pol* genes, a cDNA product was obtained that had properties consistent with those of a *pol* mRNA. Sequencing showed that the major 5' splice site (5'ss; Fig. 5) used for synthesis of *env* and *tas* mRNAs was joined to a region near the carboxyl end of the *gag* gene. Second, an independent biochemical assay, primer extension, showed that this RNA is present in infected cells. Third, and most convincing, mutation of the 3'ss in a manner that conserved the Gag amino acid sequence led to virus particles lacking Pol protein. Furthermore, insertion of an extra stop codon at the end of Gag, which would prevent ribosomal frameshifting,

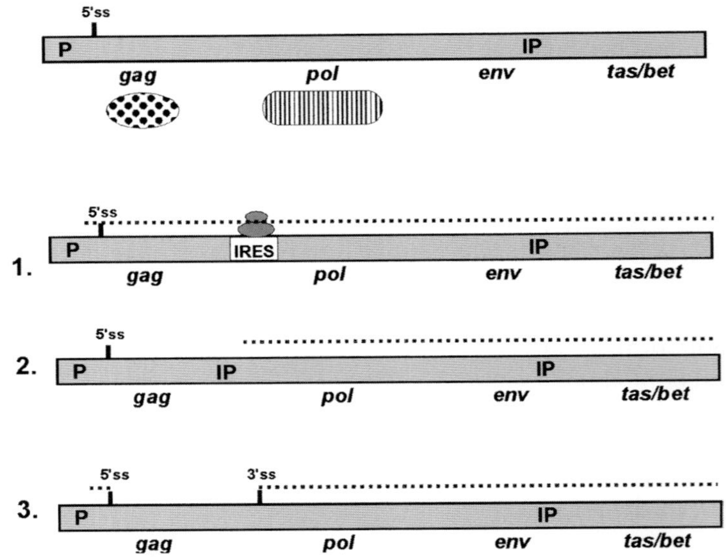

Fig. 5. Possible mechanism for synthesis of FV Pol. *1* The Pol mRNA is indicated by the *dotted line* in each case. The major 5' splice site (5''ss) is indicated. *1* Use of an internal ribosome entry site (IRES). *2* Use of an internal promoter (IP). *3* Synthesis of a Pol specific spliced mRNA using a 3' splice site (3'ss) in *gag*

did not abrogate synthesis of Pol. Similar results were obtained for PFV by other groups (Enssle et al. 1996; Jordan et al. 1996; Löchelt et al. 1996), as well as for FFV (Bodem et al. 1996), BFV (Holzschu et al. 1998), and EFV (Lecellier et al. 2002). Thus the FV Pol protein is uniquely synthesized from a spliced mRNA (Fig. 5, pathway 3). This pathway is also different from that of the hepadnaviruses such as HBV. HBV also synthesizes its reverse transcriptase protein (called P) independently from its core protein. However, in this case, P is synthesized by an internal initiation event from a full-length genomic RNA (reviewed in Ganem 1996).

2.3 Foamy Virus Pol Packaging

Because of the unique aspects of FV Pol synthesis, the pathway by which it is incorporated into particles is not shared with any other viruses. It

Table 2. Effects of PFV mutations on RNA and Pol incorporation into extracellular virions

Region of genome	Specific mutation	RNA incorporation	Pol incorporation	Protease activity	Reference
PBS-leader	△350–444[a]	low	nt	low to wt	Heinkelein et al. 2000b
U5	△317–345[a]	wt	nt	—	Heinkelein et al. 2000b
U5-PBS-leader	△194–444[a]	wt	nt	—	Heinkelein et al. 2000b
Gag	△aa 596–648[b]	wt	wt	—[c]	D. Baldwin and M. Linial, unpublished results; Baldwin et al. 1999
Gag	△aa 536–648[b]	low	—	—[c]	D. Baldwin and M. Linial, unpublished results
Pol	△2617–3870[a]	wt	nt	wt	Heinkelein et al. 1998
Pol	△3870–5773[a]	low	nt	—	Heinkelein et al. 1998

nt, Not tested; wt, approximately wild-type levels; low, up to 10-fold reduction in RNA
[a] Deletion of RNA
[b] Stop codon introduced; protein is deleted but RNA is intact
[c] Lacks protease cleavage site in Gag

has been shown that Pol incorporation is not required for RNA packaging (Baldwin et al. 1998); however, little is known about the actual requirements for Pol packaging. A mutant with a small deletion of 29 nt in U5 packaged RNA efficiently, but particles were composed solely of the full-length Gag protein, indicating that protease cleavage did not occur (Table 2, 317–345). Thus the region at nt 317–345 appears to be required for either Pol incorporation and/or protease activation (Heinkelein et al. 2000b). These authors did not directly measure Pol incorporation into particles in these experiments. Recent data provide strong evidence that viral RNA is required for Pol incorporation into particles. The authors used a four-plasmid vector system to show that Pol is not packaged unless an RNA construct with *cis*-acting sequences (see below) is present (Heinkelein et al. 2002). Several Gag mutations have been tested for their ability to incorporate Pol. A mutation that deleted both GR box I and II (aa 536–648, Table 2) failed to incorporate Pol as measured with an assay to detect Pol that had been tagged with a PKA phosphorylation site (Baldwin et al. 1999; D. Baldwin and M. Linial, unpublished results). However, this mutant also failed to incorporate genomic RNA efficiently, so whether domains within Gag itself are important for Pol incorporation is still unknown. The currently available data do seem to indicate that there is a region of RNA near the 5' end of the genome that is required for Pol packaging. Whether there is a role for Gag in Pol packaging remains to be determined. It is possible that a ternary complex between Gag, Pol, and an RNA element is required. Further mutagenesis will be required to sort out the contributions of RNA and Gag to Pol packaging.

3
RNA Genome Packaging

Retroviral genomes have all the hallmarks of cellular mRNAs including m^7G 5'' caps and polyadenylated 3' ends. Most retroviral infections do not markedly affect the metabolism of the host cell. Therefore, retroviruses are faced with the problem of selecting their genomic RNA from the large pool of mature cytoplasmic mRNAs. A large body of work is focused on understanding the details of the specific encapsidation of retroviral RNA into particles. Delineation of the salient players in encapsidation has been instrumental in the development of retroviruses as vectors for the transfer of nonviral genes.

3.1 Conventional Retroviral Packaging Sequences

In all conventional retroviruses examined, a region of the RNA near the 5' end of the genome is required for efficient, specific packaging of the genome. Such a *cis*-acting packaging sequence is commonly called psi (ψ) (reviewed in Banks et al. 1997; Berkowitz et al. 1996). Some of these ψ elements are complex and difficult to define precisely, as in the case of HIV-1. For other retroviruses, such as the alpharetroviruses, a small contiguous RNA can act as a packaging signal when placed in the context of a heterologous RNA (Banks et al. 2000). Secondary structure analyses indicate that stem loops are important components of all ψ regions. The protein partner for packaging of ψ-containing RNA is the NC domain of Gag (Fig. 2A) in the context of the complete precursor protein. Both positively charged amino acids and Cys-His boxes (zinc knuckles) contribute to the interaction with ψ RNA (Berkowitz et al. 1996; De Guzman et al. 1998). Once cleaved from the Gag precursor, and in the confines of the viral particle, NC is a nonspecific binding protein that coats the RNA, with each NC contacting about 4–7 nucleotides (Karpel et al. 1987; Khan et al. 1992). Such binding probably serves to protect the RNA from nuclease degradation. Thus NC behaves as both a specific and a nonspecific RNA binding protein. Additionally, NC has other roles in retroviral replication consistent with DNA binding activity.

3.2 Foamy Virus *Cis*-Acting Sequences

Neither the location of the RNA packaging signal of FV nor the region of Gag required for packaging has been precisely defined. The most surprising finding thus far is that the packaging signal does not appear to reside at the 5' end of the genome. Recent work (Heinkelein et al. 2000b) indicates that deletion of the U5, PBS, and leader region of PFV does not have any effect on RNA packaging whereas a smaller deletion in this region does have a significant effect on packaging ($\triangle$194–444 vs. $\triangle$350–444, Table 2). This could be due to an inadvertent second site mutation or misfolding in some other region of the RNA. Because all viruses do have specific genomic sequences required for genome incorporation, it is probable that FVs have packaging sequences as well, but where does the packaging region lie? Construction of deleted genomes to study vector transfer provides a clue. Two groups (Erlwein et al. 1998; Heinkelein, et al. 1998) found that RNA sequences encoded within the *pol* gene are required for vector transfer. Heinkelein et al. (Heinkelein et al. 1998) found that a region they call CASII (for *cis*-acting sequence two) between nt 4284 and 5624 is required for RNA packaging. Although Erlwein et al. (Erlwein et al. 1998) did not examine genome incorporation per se, they also identified a *cis*-acting sequence in this region and narrowed its location to nt 4860–5539. If this central portion of *pol* is the ψ equivalent it would be another first for FVs. Recent results show that the splice donor (SD) sequence in the R region of the FV RNA is critical for optimal Gag and Pol protein expression (Heinkelein et al. 2002). This sequence might modify the fate of newly synthesized RNA to allow efficient translation (Russell et al. 2001).

3.3 Foamy Virus *Trans*-Acting Sequences

The FV Gag proteins do not contain the hallmarks of the retroviral NC proteins, one or two copies of the Cys-His box $CX_2CX4HX4C$ embedded in regions high in positively charged amino acids, which are important determinants of recognition of genomic RNA (Table 1). It is important to note that hepadnaviruses also package RNA, which is converted to DNA before viral budding. The hepadnaviral equivalent of the Gag protein is core protein, which unlike conventional retroviral protein is not cleaved by protease. The carboxyl terminus of core protein, which is oriented toward the inside of the viral particle (Bottcher et al. 1997; Wynne et al. 1999), contains a large number of arginine residues, which are responsible for binding to both RNA and DNA of HBV (Hatton et al. 1992; Nassal 1992). The FV Gag proteins also have arginine-rich regions (GR boxes) near the carboxyl-terminal end, and these arginines may also play an important role in FV RNA packaging and interaction with the virion genome. In fact, GR box I has been shown to be important for binding to both RNA and DNA in vitro (Yu et al. 1996b). Whether or not it plays a direct role in RNA packaging is unknown; however, deletion of Box I completely eliminates viral infectivity. Boxes II and III are unlikely to be important for RNA packaging, because deletion of either box does not markedly decrease viral replication (Yu et al. 1996b) although this does not preclude a redundant signal in these boxes. In fact, deletion of both GR box II and box III does decrease RNA packaging levels (D. Baldwin and M.L. Linial, unpublished results).

4 Conclusions

Salient features of the FV particle, compared to those of conventional retroviruses and hepadnaviruses, are given in Table 3. The FV is essentially a retrovirus in hepadnaviral clothing. Evidence for a DNA genome in PFV is discussed in the chapter by Rethwilm, this volume. It is important to note that the structure of the FV Gag protein is more consistent with that of a DNA virus than that of an RNA virus. The FV Gag protein is probably multifunctional, like that of hepadnaviral core, with the amino terminus interacting with the viral glycoproteins and the carboxyl terminus mediating encapsidation of nucleic acid. It is likely that the amino terminus also targets the Gag proteins to a specific location in the cell to allow concentration and capsid formation by interacting with a cellular component. FVs are distinguished from most conventional retroviruses by lack of a condensed core in the mature virions, because of the lack of complete proteolytic processing of Gag. The role of the single cleavage in Gag is unknown, but infectious virions appear to have approximately equal amounts of cleaved and uncleaved Gag. Whether there are heterodimers between the two forms of Gag is not known. This partial cleavage of Gag is different from both retroviruses and hepadnaviruses. Overall, the organization of the FV particle is closer to that of HBV than it is to that of HIV. This is mirrored by the strict requirement for glycoprotein expression (Env or S proteins) for viral budding by FV and HBV, but not HIV. Furthermore, HBV and the primate and feline FV glycoproteins have ER retention signals that are involved in viral budding through the ER membrane (see the chapter by Lindemann and Goepfert, this volume). The FV Pol protein is very similar to those of other retroviruses in structure and activity. There are major differences, however, especially in the mode of Pol synthesis and assembly, which is unique to FVs and poorly understood. There are many challenges ahead to fully understand the regulation of the FV assembly pathways.

Table 3. Comparison of the properties of virions of retroviruses, foamy viruses, and hepadnaviruses

Viral component	Retroviruses	Foamy viruses	Hepadnaviruses
Genome	Two ss RNAs	dsDNA?	Circular gapped ds DNA
Gag	Processed to MA, CA, NC	One cleavage at carboxyl terminus	Core protein uncleaved
Pol	Packaged as Gag-Pol; processed to PR + RT-IN	Packaged without Gag determinants; processed to PR-RT + IN	Covalently bound to cDNA product; not processed
Env	Not required for budding	Required for budding	S proteins required for budding

References

Baldwin DN, Linial ML (1998) The roles of Pol and Env in the assembly pathway of human foamy virus. J Virol 72:3658–3665
Baldwin DN, Linial ML (1999) Proteolytic activity, the carboxyl terminus of Gag, and the primer binding site are not required for Pol incorporation into foamy virus particles. J Virol 73:6387–6393
Banks JD, Beemon KL, Linial ML (1997) RNA regulatory elements in the genomes of simple retroviruses. Sem Virol 8:194–204
Banks JD, Linial ML (2000) Secondary structure analysis of a minimal avian leukosis-sarcoma virus packaging signal. J Virol 74:464-
Bennett RP, Nelle TD, Wills JW (1993) Functional chimeras of the Rous sarcoma virus and human immunodeficiency virus Gag proteins. J Virol 67:6487–6498
Berkowitz R, Fisher J, Goff SP (1996) RNA packaging. Curr Top Microbiol Immunol 214:177–218
Bodem J, Löchelt M, Winkler I, Flower RP, Delius H, Flügel RM (1996) Characterization of the spliced pol transcript of feline foamy virus—the splice acceptor site of the pol transcript is located in gag of foamy viruses. J Virol 70:9024–9027
Bottcher B, Wynne SA, Crowther RA (1997) Determination of the fold of the core protein of hepatitis B virus by electron cryomicroscopy. Nature 386:88–91
Bowzard JB, Bennett RP, Krisina NK, Ernst SM, Rein A, Wills JW (1998) Importance of basic residues in the nucleocapsid sequence for retrovirus gag assembly and complementation rescue. J Virol 72:9034–9044
Campbell S, Vogt SM (1995) Self-assembly in vitro of purified CA-NC proteins from Rous sarcoma virus and human immunodeficiency virus type 1. J Virol 69:6487–6497
Cimarelli A, Luban J (2000) Human immunodeficiency virus type 1 virion density is not determined by nucleocapsid basic residues. J Virol 74:6734–6740
Craven RC, Leure-duPree RE, Weldon RA Jr, Wills JW (1995) Genetic analysis of the major homology region of the Rous sarcoma virus Gag protein. J Virol 69:4213–4227
De Guzman RN, Wu ZR, Stalling CC, Pappalardo L, Borer PN, Summers. FM (1998) Structure of the HIV-1 nucleocapsid protein bound to the SL3 psi-RNA recognition element. Science 279:384–388
Dermott E, Clarke JK, Samuels J (1971) The morphogenesis and classification of bovine syncytial virus. J Gen Virol 12: 105–119
Eastman SW, Linial ML (2001) Identification of a conserved residue of foamy virus Gag required for intracellular capsid assembly. J Virol 75:6857–6864
Enssle J, Jordan I, Mauer B, Rethwilm A (1996) Foamy virus reverse transcriptase is expressed independently from the Gag protein. Proc Natl Acad Sci USA 93:4137–4141
Erlwein O, Bieniasz PD, McClure MO (1998) Sequences in pol are required for transfer of human foamy virus-based vectors. J Virol 72:5510–5516
Fischer N, Heinkelein M, Lindemann D, Enssle J, Baum C, Werder E, Zentgraf H, Müller JG, Rethwilm A (1998) Foamy virus particle formation. J Virol 72:1610–1615

Ganem D (1996) Hepadnaviridae and their replication. In: Fields BN, Knipe DM, Howley PM (eds) Fields Virology. Lippincott-Raven, Philadelphia

Garnier L, Parent LJ, Rovinski B, Cao SX, Wills JW (1999) Identification of retroviral late domains as determinants of particle size. J Virol 73:2309–2320

Garrus JE, von Schwedler UK, Pornillos OW, Morham SG, Zavitz KH, Wang HE, Wettstein DA,.Stray KM, Cote M, Rich RL, Myszka DG, Sundquist WI (2001) Tsg101 and the vacuolar protein sorting pathway are essential for HIV-1 budding. Cell 107:55–65

Goepfert PA, Shaw KE, Bansal A, Wang G, Edwards RH, Mulligan MJ (1999) An endoplasmic reticulum retrieval signal partitions human foamy virus maturation to intracytoplasmic membranes. J Virol 73:7210–7217

Goepfert PA, Shaw KL, Ritter GD, Mulligan MJ (1997) A sorting motif localizes the foamy virus glycoprotein to the endoplasmic reticulum. J Virol 71:778–784

Hatton T, Zhou S, Standring DN (1992) RNA- and DNA- binding activities in hepatitis B virus capsid protein: a model for their roles in virus replication. J Virol 66:5232–5241

Heinkelein M, Pietschmann T, Jarmy G, Dressler M, Imrich H, Thurow J, Lindemann D, Bock M, Moebes A, Roy J, Herchenröder O, Rethwilm A (2000a) Efficient intracellular retrotransposition of an exogenous primate retrovirus genome. EMBO J 19:3436–3445

Heinkelein M, Leurs C, Rammling M, Peters K, Hanenberg H, Rethwilm A (2002) Pregenomic RNA is required for efficient incorporation of Pol polyprotein into foamy virus capsids. J Virol 76 (in press)

Heinkelein M, Schmidt M, Fischer N, Moebes N, Lindemann D, Enssle J, Rethwilm A (1998) Characterization of a cis-acting sequence in the pol region required to transfer human foamy virus vectors. J Virol 72:6307–6314

Heinkelein M, Thurow J, Dressler M, Imrich H, Neumann-Haefelin D, McClure MO, Rethwilm A (2000b) Complex effects of deletions in the 5' untranslated region of primate foamy virus on viral gene expression and RNA packaging. J Virol 74:3141–3148

Holzschu DL, Delaney MA, Renshaw RW, Casey JW (1998) The nucleotide sequence and spliced pol mRNA levels of the nonprimate spumavirus bovine foamy virus. J Virol 72:2177–2182

Jacks T, (1990) Translational suppression in gene expression in retroviruses and retrotransposons. Curr Top Micrbiol Immunol 157:93–124

Jordan I, Enssle J, Güttler E, Mauer B, Rethwilm A (1996) Expression of human foamy virus reverse transcriptase involves a spliced pol mRNA. Virology 224:314–319

Karpel RL, Henderson LE, Oroszlan S (1987) Interactions of retroviral structural proteins with single-stranded nucleic acids. J Biol Chem 262:4961–4967

Khan R, Giedroc DP (1992) Recombinant human immunodeficiency virus type 1 nucleocapsid (NCp7) protein unwinds tRNA. J Biol Chem 267:6689–6695

Konvalinka J, Löchelt M, Zentgraf H, Flügel RM, Kräusslich H-G (1995) Active spumavirus proteinase is essential for virus infectivity but not for formation of the Pol polyprotein. J Virol 69:7264–7268

Lecellier CH, Neves M, Giron M-L, Tobaly-Tapiero J, Saïb A (2002) Further characterization of equine foamy virus reveals unusual features among the foamy viruses. J Virol 76:7220–7227

Lindemann D, Bock M, Schweizer M, Rethwilm A (1997) Efficient pseudotyping of murine leukemia virus particles with chimeric human foamy virus envelope proteins. J Virol 71:4815–4820

Lindemann D, Pietschmann T, Picard-Mareau M, Berg A, Heinkelein M, Thurow J, Knaus P, Zentgraf H, Rethwilm A (2001) A particle associated glycoprotein signal peptide essential for virus maturation and infectivity. J Virol 75:5762–5771

Löchelt M, Flügel RM (1996) The human foamy virus pol gene is expressed as a Pro-Pol polyprotein and not as a Gag-Pol fusion protein. J Virol 70:1033–1040

Mammano F, Ohagen A, Hoglund S, Göttlinger HG (1994) Role of the major homology region of human immunodeficiency virus type 1 in virion morphogenesis. J Virol 68:4927–4936

Martin-Serrano J, Zang T, Bieniasz PD (2001) HIV-1 and Ebola virus encode small peptide motifs that recruit Tsg101 to sites of particle assembly to facilitate egress. Nature Med 7:1313–1319

Meiering CD, Comstock KE, Linial ML (2000) Human foamy virus multiple integrations in persistently infected human erythroleukemia cells. J Virol 74:1718–1726

Nassal M (1992) The arginine-rich domain of the hepatitis B virus core protein is required for pregenome encapsidation and productive viral positive-strand DNA synthesis but not for virus assembly. J Virol 66:4107–4116

Petit C, Giron M-L, Tobaly-Tapiero J, Bittoun P, Real E, Jacob Y, Tordo N, de Thé H, Saïb A (2003) Targeting of incoming retroviral Gag to the centrosome involves a direct interaction with dynein LC8. J Virol (in press)

Raux H, Flamand A, and Blondel D (2000) Interaction of the rabies virus P protein with the LC8 dynein light chain. J Virol 74:10212–10216

Rein A. McClure MR, Rice NR, Luftig RB, Schultz AM (1986) Myristylation site in Pr65gag is essential for virus particle formation by Moloney murine leukemia virus. Proc Natl Acad Sci USA 83:7246–7250

Renshaw RW, Casey JW (1994) Transcriptional mapping of the 3' end of the bovine syncytial virus genome. J Virol 68: 1021–1028

Rhee SS, Hunter E (1987) Myristylation is required for intracellular transport but not for assembly of D-type retrovirus capsids. J Virol 61:1045–1053

Rhee SS, Hunter E (1990) A single amino acid substitution within the matrix protein of a type D retrovirus converts its morphogenesis to that of a type C retrovirus. Cell 63:77–86

Russell RA, Zeng Y, Erlwein O, Cullen BR, McClure MO (2001) The R region found in the human foamy virus long terminal repeat is critical for both gag and pol protein expression. J Virol 75:6817–6824

Saïb A, Puvion-Dutilleul F, Schmid M, Peries J, de Thé H (1997) Nuclear targeting of incoming human foamy virus gag proteins involves a centriolar step. J Virol 71:1155–1161

Sakalian M, Hunter E (1999) Separate assembly and transport domains within the Gag precursor of Mason-Pfizer monkey virus. J Virol 73:8073–8082

Sandefur S, Smith RM, Varthakavi V, Spearman P (2000) Mapping and characterization of the N-terminal I domain of human immunodeficiency virus Type 1 Pr55(Gag). J Virol 74:7238–7249

Schliephake AW, Rethwilm A (1994) Nuclear localization of foamy virus Gag precursor protein. J Virol 68:4946–4954

Spearman P, Wang JJ, van der Heyden N, Ratner L (1994) Identification of human immunodeficiency virus type 1 Gag protein domains essential to membrane binding and particle assembly. J Virol 68:3232–3242

Swanstrom R, Wills JW (1997) Synthesis, assembly and processing of viral proteins. In: Coffin JM, Hughes SH, Varmus HE (eds) Retroviruses. Cold Spring Harbor Laboratory Press, New York

Tobaly-Tapiero J, Bittoun P, Neves M, Guillemin MC, Lecellier CH, Puvion-Dutilleul F, Gicquel B, Zientara S, Giron M-L, de Thé H, Saïb A (2000) Isolation and characterization of an equine foamy virus. J Virol 74:4064–4073

VerPlank L, Bouamr F, LaGrassa TJ, Agresta B, Kikonyogo A, Leis J, Carter CA (2001) Tsg101, a homologue of ubiquitin-conjugating (E2) enzymes, binds the L domain in HIV type 1 Pr55(Gag). Proc Natl Acad Sci USA 98:7724–7729

Wilk T, Geiselhart V, Frech M, Fuller SD, Flügel RM, Löchelt M (2001) Specific interaction of a novel foamy virus Env leader protein with the N-terminal Gag domain. J Virol 75:7995–8007

Wynne SA, Crowther RA, Leslie AG (1999) The crystal structure of the human hepatitis B virus capsid. Mol Cell 3:771–780

Yu SF, Baldwin DN, Gwynn SR, Yendapalli S, Linial ML (1996a) Human foamy virus replication—a pathway distinct from that of retroviruses and hepadnaviruses. Science 271:1579–1582

Yu SF, Edelmann K, Strong RK, Moebes A, Rethwilm A, Linial ML (1996b) The carboxyl terminus of the human foamy virus gag protein contains separable nucleic acid binding and nuclear transport domains. J Virol 70:8255–8262

Yu SF, Linial ML (1993) Analysis of the role of the bel and bet open reading frames of human foamy virus by using a new quantitative assay. J Virol 67:6618–6624

Yuan B, Campbell S, Bacharach E, Rein A, Goff SP (2000) Infectivity of moloney murine leukemia virus defective in late assembly events is restored by late assembly domains of other retroviruses. J Virol 74:7250–7260

Chapter 5

The Foamy Virus Envelope Glycoproteins

D. Lindemann, P. A. Goepfert

Abstract The main functions of retroviral glycoproteins are recognition and binding to the cellular virus receptor as well as fusion of viral and cellular lipid membranes to release the viral particle into the cytoplasm of the host cell. Foamy viruses (FVs) are a special group of retroviruses with a very broad host range that use a currently unknown cellular receptor for entry. Nevertheless, many functions of the FV envelope glycoproteins in the viral replication cycle have been characterized in detail over the last years. Several unique features not found for any other retro-

D. Lindemann (✉)
Institut für Virologie, Medizinische Fakultät Carl-Gustav-Carus,
Technische Universität Dresden,
Fetscherstraße 74, 01307 Dresden, Germany
E-mail: dirk.lindemann@mailbox.tu-dresden.de

P. A. Goepfert
Division of Infectious Diseases, Department of Medicine,
University of Alabama at Birmingham,
220 Bevill Biomedical Research Building, 845 19th St. South, Birmingham,
AL 35294, USA

virus were identified. These include the presence of two types of FV Env proteins, gp170$^{Env-Bet}$ and gp130^{Env}, and the strict requirement of gp130^{Env} coexpression for the FV budding and particle release process, a function that cannot be compensated for by any other viral glycoprotein tested so far. Furthermore, domains in gp130^{Env} could be characterized that influence its intracellular distribution, cell surface transport, and its specific interaction with the viral capsid during particle egress. In addition, it has recently been shown that gp130^{Env} expression alone induces release of subviral particles from cells. This review summarizes the current knowledge about the nature of the FV Env proteins and their function in the viral replication cycle.

1 Introduction

The essential functions of glycoproteins of enveloped viruses are attachment of the viral particle to target cells by binding to cellular receptors and subsequent fusion of viral and cellular lipid membranes to release the viral capsid into the cytoplasm (reviewed in Flint et al. 2000). The entry of influenza virus from the endosomal pathway involving the hemagglutinin (HA) glycoprotein is one of the best-understood viral entry mechanism pathways (Skehel and Wiley 2000). The fusion reaction mediated by the influenza HA and its structure show remarkable similarities to the envelope glycoproteins of two retroviruses, the human immunodeficiency virus (HIV) and the murine leukemia virus (MuLV), which are the best studied retroviral envelope proteins to date (reviewed in Hunter 1997).

Similar to orthoretroviruses, foamy viruses (FVs) encode for a type I membrane glycoprotein; however, recent analysis of the structure and function of the FV Env proteins, has revealed several unique features among these glycoproteins. The focus of this review is to summarize the current knowledge about the nature of the FV Env proteins and their function in the FV replication cycle. If not otherwise stated, this synopsis concentrates on the Env protein of the prototype foamy virus (PFV) isolate that has been studied most extensively so far. If there are experimental data of FV glycoproteins from other species, where deviations in respect to the PFV Env are observed, they are discussed accordingly.

2
The FV Env Proteins

Many enveloped viral particles contain separate attachment and fusion proteins (reviewed in Flint et al. 2000). In contrast, retroviral glycoproteins are normally synthesized as single precursor proteins that mediate both functions. They are cleaved posttranslationally by cellular proteases during their transport to the plasma membrane into a surface (SU) and a transmembrane (TM) subunit. The SU subunit is primarily involved in binding to the viral receptor on the host cell, whereas the TM subunit is required for fusion of viral and cellular lipid membranes to release the viral capsid into the cytoplasm of the host cells (reviewed in Hunter 1997).

The SU glycoprotein of most retroviruses binds to a specific cellular receptor and is one of the factors that limit the type of cell that can be infected by a particular retrovirus. The most notable example of this phenomenon is seen with HIV-1 infection, where only cells expressing both CD4 and a chemokine receptor are efficiently infected (Deng et al. 1996; Dragic et al. 1996; Feng et al. 1996; Samson et al. 1996). Because FVs have an extremely broad cellular tropism (see Falcone et al., this volume), it is believed that either the FV receptor is expressed by a large number of cells or the FV surface glycoprotein is able to bind several different cellular receptors. Although virtually nothing is known about this topic, defining the FV cellular receptor will be important if these viruses are to be used as vectors for gene delivery in the future.

Unlike what has been reported for other retroviruses, two species of Env precursor glycoproteins with approximate sizes of 130 and 170 kDa are detected in lysates of FV-infected cells, which are expressed approximately at a ratio of 3 to 1, respectively (Fig. 1; Giron et al. 1998, 1993; Lindemann and Rethwilm 1998; Netzer et al. 1990). In addition, cleavage products of 110 and 80–90 kDa derived from gp170 or 80–90 and 48 kDa derived from gp130 are detected by Env specific antisera (Fig. 1).

The 170-kDa precursor was shown to be a chimeric protein containing sequences of the env ORF and an accessory ORF, whereas the 130-kDa form represents the precursor protein of the particle-associated FV Env protein translated exclusively from the env ORF. Both proteins and their respective cleavage products are discussed below. However, because only gp130 is essential for FV replication in vitro and more information

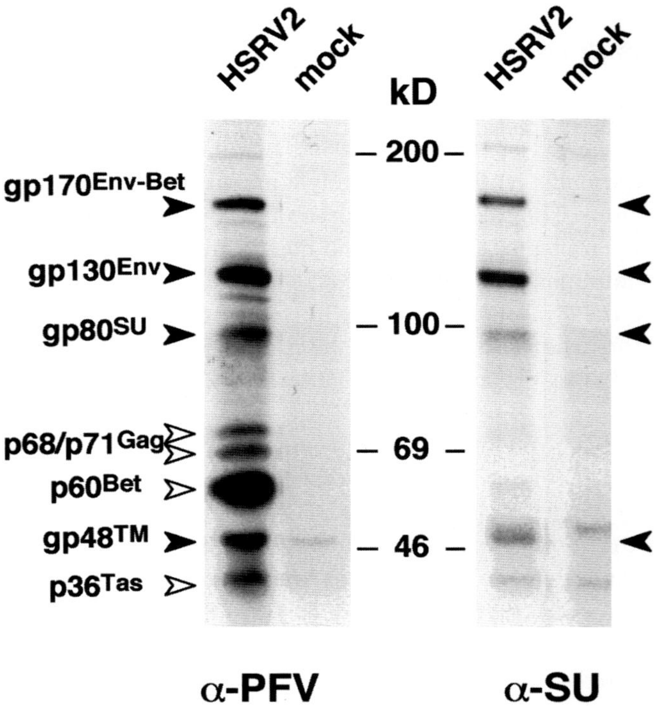

Fig. 1. Radioimmunoprecipitation analysis of 293T cells transfected with a PFV proviral expression clone. 293T cells were transfected with proviral expression clone pcHSRV2, and cell lysates were immunoprecipitated with a chimpanzee serum recognizing all major PFV proteins (as indicated) as described by Lindemann and Rethwilm (1998)

about its function and role in the FV replication cycle is available, it is the main topic of this review.

2.1 The Env-Bet Fusion Protein

With site-directed mutagenesis approaches combined with immunodetection employing FV Env-, Bet-, or Tas-specific antisera it has been shown that the 170-kDa glycoprotein is an Env-Bet fusion protein (Giron et al. 1998; Lindemann and Rethwilm 1998). It is expressed from an alternatively spliced Env mRNA utilizing a splice donor-acceptor pair lo-

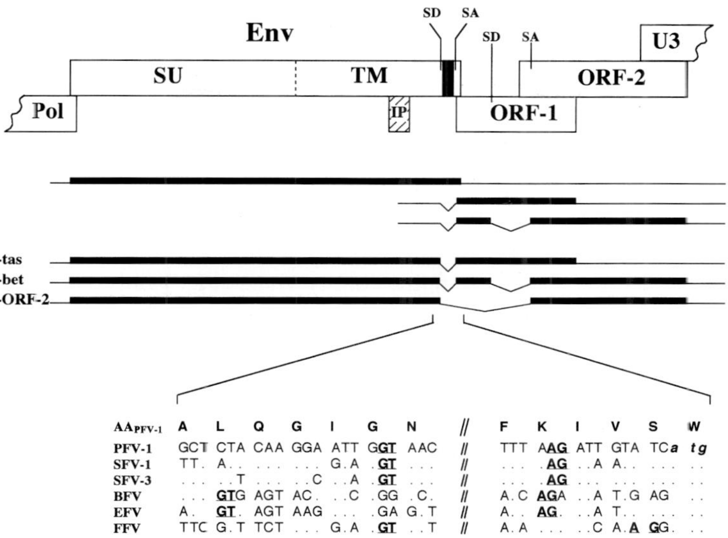

Fig. 2. Schematic illustration of the env and accessory ORFs and the corresponding mRNAs. The different ORFs are indicated as *boxes* and the coding capacities of the individual mRNAs as *thick lines*. The nucleotide sequence of FV isolates from various species involved in the alternative splicing event leading to Env-Bet expression is shown *below* with the SD and SA sites used indicated in *bold*. *SD*, splice donor; *SA*, splice acceptor; *IP*, internal promoter

cated downstream of the internal promoter within the env ORF (Giron et al. 1998; Lindemann and Rethwilm 1998; Fig. 2). Furthermore, the splice donor-acceptor pair is conserved among all known FVs. Normally this pair is used to generate spliced tas and bet transcripts derived from the internal promoter (Fig. 2; Löchelt et al. 1994; Mergia 1994). The Env-Bet protein consists of the extracellular domains of the gp130^{Env} fused to the full-length Bet protein with three amino acids in between not found in either wild-type protein but lacks the membrane-spanning (MSD) and cytoplasmic (CyD) domains of the TM subunit. The gp170$^{Env\text{-}Bet}$ precursor, as well as its cleavage products gp110$^{TM\text{-}Bet}$ and gp80^{SU}, are detected in cell lysates and in the supernatant of FV-infected cells (Giron et al. 1998; Lindemann and Rethwilm 1998). However, they are not particle associated (Lindemann and Rethwilm 1998) and only the Bet protein itself, which is also found in the supernatant, is taken up by neighboring cells (Giron et al. 1998). Although an env-bet mRNA could be PCR amplified from infected cells in both studies, an env-tas mRNA could only be PCR amplified by one of the groups (Giron et al. 1998). In contrast, the Env-Tas protein was undetectable in both studies undertaken (Giron et al. 1998; Lindemann and Rethwilm 1998). Although theoretically possible, a PFV-Env-ORF-2 protein was also undetectable at protein and mRNA levels. Similar to the primate FVs, the equine foamy virus (EFV) can theoretically generate an Env-Bet protein (Fig. 2) and the corresponding mRNA is indeed detectable (Lecellier et al., personal communication). Interestingly, for feline foamy virus (FFV) only an Env-ORF-2 protein could be detected (Bodem et al. 1998). However, this is not surprising, because a potential env-bet mRNA results in an out of frame fusion of env ORF and ORF-1, and only splicing of env ORF and ORF-2 allows the expression of a fusion protein (Fig. 2). This might indicate that FFV sequences encoded by the ORF-2 and not by the ORF-1 (in contrast to PFV) may serve important functions for the fusion proteins.

Although the nature and expression mechanism of the Env-Bet protein have been characterized in great detail, the function of the Env-Bet protein in the replication cycle of FV remains unknown. Recombinant FV lacking an Env-Bet protein, through mutational inactivation of the conserved splice donor-acceptor pair within the env ORF, replicates in vitro with kinetics comparable to that of wild-type virus (Lindemann and Rethwilm 1998). Furthermore, a reversal of the introduced mutations is not observed on long-term cultivation, implying that there is no in vitro selection pressure for retaining Env-Bet expression (Lindemann and Rethwilm 1998). Similar experiments for EFV Env-Bet or FFV Env-ORF-2 have not yet been reported. Although the described experiments suggest that the Env-Bet protein may play a role in viral replication or infectivity in vivo, these studies have yet to be performed.

2.2
Intracellular Distribution of Gp130^{Env}, Sorting and Transport Signals

Gp130^{Env} is the precursor of the particle-associated glycoprotein subunits gp80^{SU} and gp48^{TM} that are generated by a cleavage event involving a cellular protease during the intracellular transport of the precursor (Fig. 3; Goepfert et al. 1997; Lindemann et al. 1997). Precursor processing is strongly reduced in the absence of coexpression of other FV proteins, and the majority of the Env protein is found in the endoplasmic reticulum (ER), as determined by immunofluorescence and endoglyco-

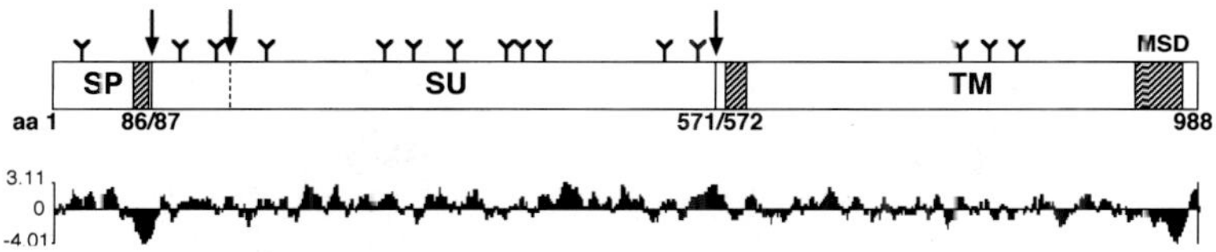

Fig. 3. Schematic illustration of the gp130^{Env} protein. The gp130^{Env} protein is illustrated as a *box*. The predicted cleavage sites leading to subunit processing are indicated. The hydrophobic sequences of the signal peptide (*LP*), the fusion peptide, and the membrane-spanning domain are drawn as *shaded boxes*. The positions of potential N-glycosylation sites are shown as *Y*. Below a hydrophilicity plot of the gp130^{Env} protein sequence is given

sidase H analysis (Goepfert et al. 1997; Pietschmann et al. 1999). This characteristic is attributed at least in part to a special structural feature of the FV Env protein, a dilysine ER retrieval signal found at the C-terminus of the TM cytoplasmic domain (Goepfert et al. 1995, 1997). It clearly influences the intracellular distribution of gp130^{Env}, as determined by endoglycosidase H analysis (Goepfert et al. 1997; Pietschmann et al. 1999); however, there are conflicting results regarding its role for the Env cell surface transport. Expression of ER retrieval signal mutants in COS-1 cells has been reported to result in increased syncytia formation compared to the wild-type protein (Goepfert et al. 1997). A more quantitative analysis of the fusion activity involving 293T and HT1080 cells and cell surface biotinylation analysis of one of the ER retrieval signal mutants, on the other hand, failed to confirm these data and revealed only minor differences compared to the wild-type glycoprotein (Pietschmann et al. 2000). It is thought that the ER retrieval signal is at least partially responsible for FV budding predominantly into the ER, which is supported by the finding that ER retrieval mutants show an enhanced plasma membrane budding (Goepfert et al. 1999). However, these mutants do not give rise to higher extracellular viral titers, and, furthermore, the ER retrieval signal and even the complete CyD of the TM subunit are dispensable for viral infectivity in vitro (Goepfert et al. 1999; Pietschmann et al. 1999). Additionally, mutants of the ER retrieval signal did not revert to wild type despite prolonged passage in cell culture (Goepfert et al. 1999). Interestingly, the Env protein of the recently cloned EFV isolate seems to lack a functional ER retrieval signal and budding of EFV exclusively at the plasma membrane has been reported (Tobaly-Tapiero et al. 2000). Although the significance of the ER retriev-

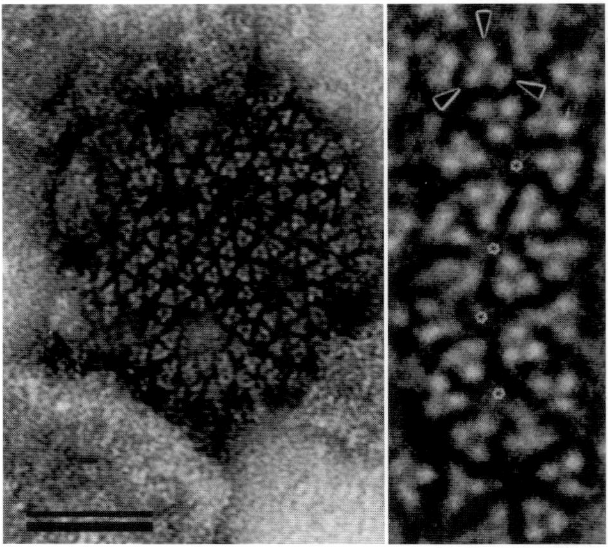

Fig. 4. Surface features of PFV particles detected by negative-staining EM. PFV displays a network of trimeric viral spike proteins (*left panel*), predominantly arranged into rings of six subunits. When grouped in hexameric rings, a stain-filled hole with a diameter of about 8 nm is formed (*right panel, asterisks*). Adjacent rings share two completely integrated spikes. Images at higher magnification reveal three separate densities (*arrowheads*) in the triangular spike (*right panel*). *Bar* represents 50 nm (*left panel*). (From T. Wilk and M. Löchelt)

al signal for FV replication in vitro is currently unclear, this highly conserved amino acid sequence (at least in primate FVs) suggests that it plays an important role for the replication of these viruses in vivo.

At least two additional regions of the FV Env protein influence the cell surface expression as well. One is the MSD of the TM subunit, in particular a lysine residue in the middle of the putative domain found conserved in all FV isolates, which seems to inhibit cell surface transport (Pietschmann et al. 2000). Deletion or C-terminal truncation of the MSD and various mutations of the conserved lysine even in combination with a wild-type ER retrieval signal result in increased cell surface Env expression (Pietschmann et al. 2000). The positive charge at this position seems to be the critical determinant, because a conservative $K_{959}R$ exchange mutant displays a phenotype similar to the wild-type protein. The cytoplasmic N-terminal leader peptide (LP) sequence of the FV Env

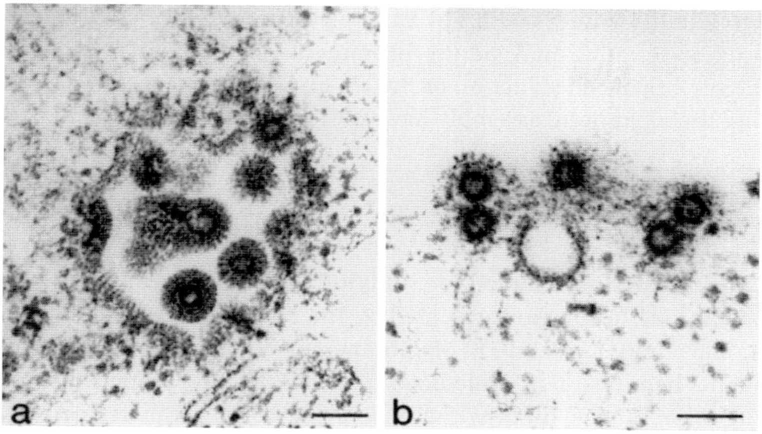

Fig. 5a, b. Electron micrographs showing representative negative-stained thin sections of 293T cells transfected with a PFV proviral clone. Budding of PFV is observed predominantly at intracellular membranes (**a**), but to a lesser extent also at the plasma membrane (**b**). *Bar* represents 100 nm. (From H. Zentgraf)

protein is a second domain responsible for intracellular retention of FV Env in the absence of coexpression of other FV proteins (Lindemann et al. 2001). N-terminal truncation of the LP leads to a dramatic increase of Env detectable at the cell surface, and this result occurred even with a wild-type ER retrieval signal present in the cytoplasmic domain of TM (Lindemann et al. 2001), similar to what was observed for the MSD mutants (Pietschmann et al. 2000). In contrast to the ER retrieval signal neither the MSD nor the LP is dispensable for FV infectivity (Goepfert et al. 1999; Lindemann et al. 2001; Pietschmann et al. 1999), but it might be possible that an interaction of all three regions is required for correct intracellular distribution of the FV Env protein.

Although FV gp130^{Env} subunits form prominent spike structures on budding and secreted FV particles, as seen with electron microscopy (EM) analysis (Figs. 4 and 5), little is known about its oligomeric organization or three-dimensional structure. One report using three-dimensional image reconstruction analysis of EM pictures (Fig. 4) indicates that the gp130^{Env} subunits are organized in trimers in the lipid membrane (Wilk et al. 2000), similar to what was reported for the MuLV and HIV glycoproteins (Chan et al. 1997; Fass et al. 1996, 1997). Interestingly, this represents the only study involving retroviral glycoproteins that was able to identify trimers with EM analysis alone (see Fig. 4). The reasons for this observation may be the prominent extracellular spikes and/or the highly organized structure of the FV Env. However, the domains mediating the oligomeric assembly have yet to be characterized.

3
Analysis of Envelope Fusion Activity

The most widely used model for studying how viral glycoproteins mediate membrane fusion is represented by the influenza HA (Bullough et al. 1994; Carr and Kim 1993). Membrane fusion is believed to proceed through a hemifusion intermediate in which the outer leaflets of two opposing bilayers fuse, followed by fusion of the inner leaflets and the formation of the fusion pore (Carr et al. 1997). Recent results indicate that retroviral glycoprotein-mediated membrane fusion seems to follow a similar mechanism (Chan et al. 1997; Weissenhorn et al. 1997). They also demonstrated that, in addition to the fusion peptide, the MSD and CyD domains are also essential or have important regulatory functions.

Little is currently known about the FV Env-mediated membrane fusion process. No high-resolution three-dimensional structures of the FV Env or subunits thereof are available. However, structural predictions based on comparison of the FV Env protein sequence with that of other viral glycoproteins point to homologies of the FV TM subunit to that of orthoretroviral glycoproteins such as HIV-1 or MuLV (Wang and Mulligan 1999). Similar to those glycoproteins, FV Env contains an amphipatic α-helical structure near the N-terminus of the TM subunits, probably representing the fusion peptide (FP) (Wang and Mulligan 1999). Recently, the theory that the FV FP is an amphipatic α-helix was confirmed experimentally (Wang et al., manuscript in preparation). Based on a periodicity of 3.6 amino acids for α-helices and 2.0 for β-sheets within protein structures, it is predicted that inserting two amino acids into the middle of a FP would disrupt a α-helical but not a β-sheet structure. Conversely, adding one or three amino acids into the FP would disrupt a β-sheet and not an α-helix. Insertion of two alanines into the middle of the putative FP dramatically reduced formation of syncytia by mutant $gp130^{Env}$. In contrast, one or three alanines inserted into the middle of the $gp130^{Env}$ FP had little affect on syncytia formation. Therefore, these data strongly implied that the functional conformation of the FV FP is the α-helix.

Analysis of peptides of this region by circular dichroism revealed both α-helical and β-structure conformation or predominantly α-helical conformation in aqueous solution or trifluoroethanol, respectively (Epand and Epand 2001). Interestingly, unlike several other fusion peptides studied previously, these FV peptides did not promote increase membrane curvature although promoting membrane fusion in a synergistic fashion with a peptide of the N-terminal region of the putative MSD (Epand and Epand 2001). Another recent report using sequence alignment of regions of the FV TM subunit with known fusion peptides only found similarity to those with a retro sequence (writing it in reverse from C- to N-terminus) spanning the N-terminal region of the putative FV MSD and upstream flanking sequences (Peisajovich and Shai 2002).

An additional unusual feature of the FV Env is the presence of two putative α-helix domains, one immediately downstream of the fusion peptide and a second just upstream of the MSD. These α-helices appear similar in position and length to those observed in the solved structures of influenza HA2 and HIV-1 gp41 (Wang and Mulligan 1999). An unusual structural characteristic of the FV Env TM subunit is a lengthy putative central region, containing β-sheets and loops, increasing the length of the extracellular domain of the FV TM subunit to more than twice the length of that of any orthoretrovirus (Wang and Mulligan 1999).

The limited experimental analysis of the FV Env-mediated fusion process that has been performed up to this time has revealed several features shared with orthoretroviral glycoproteins. As mentioned above, the gp130^{Env} precursor protein is cleaved by a furinlike cellular protease into SU and TM subunits (Goepfert et al. 1997). Inactivation of the cleavage site by site-directed mutagenesis revealed that envelope precursor cleavage is also a prerequisite for the FV fusion process and infectivity of FV particles (Bansal et al. 2000; Pietschmann et al. 2000). Intracellular distribution of such a cleavage mutant is comparable to the wild-type protein; however, FV particle release (discussed below) is somewhat reduced (Bansal et al. 2000; Pietschmann et al. 2000). Whereas the fusogenic potential of the HIV-1 and MuLV envelope protein is influenced by the CyD of their TM subunit, the FV CyD is dispensable for fusion activity and seems to only marginally, if at all, affect the fusogenic capacity (Pietschmann et al. 2000). However, as found for other retroviral glycoproteins, deletion of the MSD or parts thereof abolishes fusion activity and viral infectivity, showing that the full-length MSD is involved in the fusion process (Pietschmann et al. 2000). The conserved lysine found in the middle of the MSD seems to be involved in regulation of FV Env fusion activity, in addition to affecting cell surface expression, because at least one mutant displayed a dramatically increased fusion activity with cell surface expression levels comparable to the wild-type protein (Pietschmann et al. 2000).

4 Env and FV Particle Release

FV are unique among all retroviruses in respect to their viral particle assembly process. Similar to B- or D-type retroviruses, they preassemble their capsid in the cytoplasm of infected cells before budding occurs (reviewed in Swanstrom and Wills 1997). Budding of FV across cellular membranes is observed predominantly at the ER, and to a lower degree also at the plasma membrane (Fig. 5; Baldwin and Linial 1998; Fischer et al. 1998; Zemba et al. 1998). However, if only FV Gag or Gag/Pol proteins are expressed, virus-like particles cannot be detected in the cell culture supernatant and only accumulated naked capsids, which are not associated with cellular membranes, are found by electron microscopy analysis in the cytoplasm of such cells (Baldwin and Linial 1998; Fischer et al. 1998; Pietschmann et al. 1999). Therefore, gp130^{Env} is absolutely necessary for budding across cellular membranes and viral particle release to take place (Baldwin and Linial 1998; Fischer et al. 1998; Pietschmann et al. 1999). This suggests that the FV Gag protein contains all the structural information for capsid assembly whereas the FV gp130^{Env} harbors additional determinants required for the budding and viral release steps of the FV replication cycle. These findings are in contrast to orthoretroviruses, where the Gag protein by itself carries the structural information necessary for capsid assembly and budding across cellular membranes to release viruslike particles (Swanstrom and Wills 1997). Furthermore, they imply inherent specific interactions between FV Gag and Env proteins because heterologous viral glycoproteins, such as the MuLV envelope protein or the vesicular stomatitis virus glycoprotein G, are unable to replace the FV Env protein in this step of the FV replication cycle (Pietschmann et al. 1999).

Two regions of gp130^{Env} so far have been identified that are crucial for the FV budding process. The first is the putative MSD of the TM subunit. C-terminal deletion analysis of gp130^{Env} revealed that the CyD of the FV Env is dispensable for FV particle release whereas the N-terminal

15 aa of the putative FV MSD are essential (Pietschmann et al. 1999, 2000). Alternative membrane anchorage of the gp130^{Env} extracellular domains via a glycophospholipid tail is unable to restore FV particle release (Pietschmann et al. 1999). Interestingly, a chimeric mutant, in which the putative FV MSD was replaced by that of the MuLV envelope protein, shows budding into intracellular compartments; however, they are noninfectious, and particle release into the supernatant is not supported (Pietschmann et al. 1999). Furthermore, in the context of the MuLV MSD chimera, a supplementary FV CyD with a functional ER retrieval signal is required, because intracellular budding of this chimera is abolished by inactivation of the ER retrieval signal or replacement of the FV CyD by that of MuLV Env (T. Pietschmann and D. Lindemann, unpublished data).

Another unusual structural feature of FV Env proteins is a very long putative N-terminal LP domain (Fig. 3), which is a second region required for FV budding (Lindemann et al. 2001). In the case of PFV Env and FFV Env, the LPs have a predicted length of 86 aa and 83 aa, respectively, containing an extended cytoplasmic N-terminal domain (Flügel et al. 1987; Wang and Mulligan 1999). However, recent analysis with LP-specific antisera revealed that the major cleavage products of approximate sizes of 18 kDa for PFV and 16.5 kDa for FFV are larger than expected for LP with predicted length of 86 and 83 aa, respectively (Lindemann et al. 2001; Wilk et al. 2001). Mutational analysis for the PFV Env indicated that the LP cleavage site is in fact located further downstream (Lindemann et al. 2001). Furthermore, the analysis showed that cleavage by a cellular protease is predominately posttranslational (Lindemann et al. 2001). Surprisingly, LP cleavage products were found to be viral particle associated, pointing to a role for the FV Env LP in the budding process (Lindemann et al. 2001). This observation was confirmed for the FFV LP cleavage product (Wilk et al. 2001). Indeed, a so-called "budding domain" comprising the N-terminal 15 aa with two conserved tryptophan residues was characterized by genetic experiments involving deletion and point mutants as being essential for PFV budding and particle release (Lindemann et al. 2001). In contrast, the same part of the PFV Env LP was not required for targeting into the secretory pathway as mutants lacking it were transported to the cell surface and efficiently pseudotype infectious MuLV particles (Lindemann et al. 2001). In surface plasmon resonance analyses of recombinant FFV Env (aa 1–65) and FFV Gag (aa 1–54) subdomains, a direct and specific interaction was observed in vitro (Wilk et al. 2001). The important role of the two conserved tryptophan residues for this interaction was confirmed by using alanine substitution mutants in these analyses. An N-terminal 30 aa domain of FFV Env was sufficient for the interaction; however, the complete cytoplasmic domain of the FFV Env protein (aa1–65) resulted in a significantly stronger binding (Wilk et al. 2001). Furthermore, preliminary results of pull-down experiments using recombinant GST-PFV-LP fusion proteins and membrane-stripped PFV particles also point to a direct interaction between FV Env and the viral capsid (Eastman and Linial, personal communication).

Taken together, both the N-terminal part of the LP domain and the MSD are obviously required to be present simultaneously because deletion of either one interferes with normal FV budding and particle release.

A model of the FV budding process that could be considered would encompass an initial interaction of the preassembled FV capsid and the LP domain of the FV gp130^{Env} precursor protein, which is subsequently followed or stabilized by a second contact with the N-terminal region of the MSD in TM. However, to confirm this model, a more detailed analysis of the sequential steps leading to budding are required. In vitro binding studies for the FFV and PFV Env point to a direct interaction between N-terminal LP budding domain and the N-terminus of the Gag protein. However, currently nothing is known about the interaction of the viral capsid and the FV Env MSD or whether cellular proteins are involved.

As stated previously for orthoretroviruses, Gag expression alone is sufficient for assembly and release of particles. In contrast, pararetroviruses such as hepatitis B form subviral particles (SVP) that bud across plasma membranes when envelope alone is expressed in the absence of capsid proteins. Because FV Env is essential for extracellular particle budding, it seems possible that this glycoprotein could also direct budding of SVP when expressed alone in the same way as hepatitis B. Additionally, EM studies of the infectious FV particles (Wilk et al. 2000) suggest that the FV gp130^{Env} has a more ordered structure compared to the Env of orthoretroviruses, further indicating that gp130^{Env} might form SVPs. So it is perhaps not surprising that FV gp130^{Env} was recently shown to form SVP and bud across cellular membranes in the absence of any other FV protein expression (Shaw et al. 2003 in press). FV gp130^{Env} pelleted through a 20% sucrose cushion. Furthermore, this pelleted material was sensitive to treatment with 0.5% NP-40, suggesting that these particles contained lipid. They also had a density between 1.13 and 1.17 g/ml as measured by centrifugation experiments over a sucrose gradient. Interestingly, this density was only slightly less than what was seen with infectious FV. The particles formed by the FV $gp130^{Env}$ protein alone were also shown to be specific by immunogold labeling experiments as viewed by EM. Together these findings indicate that the particles formed by the FV $gp130^{Env}$ are indeed SVPs. These characteristics are reminiscent of pararetroviruses and further separate FV from orthoretroviruses. Unlike hepatitis B however, it is still not clear whether the SVPs of FV are produced in excess of infectious virions, or whether SVPs occur when FV Gag is coexpressed with $gp130^{Env}$. Furthermore, it remains to be determined whether SVPs play an important role in the FV replication cycle.

5
Conclusions

FVs are a special group of retroviruses that deviate in many aspects from the replication strategy found for orthoretroviruses. The FV Env proteins make no exception to this rule. First, there are at least two kinds of FV Env protein. A 130-kDa form resembling in its subunit organization the glycoprotein precursor of orthoretroviruses is produced, but in addition FVs express a 170-kDa fusion protein consisting of the extracellular domains of the $gp130^{Env}$ and aa encoded by the accessory ORFs. This fusion protein, being dispensable for FV replication in vitro, has been attributed no apparent function. However, it seems unlikely that a virus would express an evolutionary conserved protein at such a level without benefit for its propagation. To elucidate its role in the FV replication cycle, future in vivo experiments employing mutant viruses and analysis of their replication in an animal model are required.

The $gp130^{Env}$ shows several unique structural features, one being an ER retrieval signal in its C-terminal cytoplasmic domain that influences intracellular distribution but is nonessential for in vitro replication. Furthermore, FV Env proteins form prominent spike structures on viral particles or infected cells. Interestingly the FV TM subunit contains extended β-sheets and loops containing a putative central region. Therefore, it will be interesting to obtain a high-resolution three dimensional structure of the FV TM subunit and analyze the function of this domain for the oligomeric structural organization because extracellular TM subunit domains of other retroviral glycoproteins have been shown to be involved in this process. Three-dimensional image reconstruction analysis performed on EM pictures of FV virions indicate a trimeric organization of FV Env proteins in the spike structures, similar as reported for other retroviral glycoproteins (Wilk et al. 2000).

To date the most intriguing finding for the particle associated $gp130^{Env}$ protein is its essential role for the FV budding and particle release process. Furthermore $gp130^{Env}$ when expressed in the absence of other FV proteins forms SVPs that bud into the media. Two regions, the N-terminal domains of the putative MSD in TM and the LP have been shown to be required for budding of FV particles at cellular membranes and their release into the supernatant. It will be interesting to determine if these domains are also involved in SVP formation. Additionally, the order of events leading to budding of FV capsids across membranes or the interaction of FV Env and capsid proteins needs to be examined in further detail to get a better understanding of this process. It currently remains unknown whether the FV capsids budding predominantly into the ER are actively exported into the supernatant by the cellular secretory machinery or whether they are simply released as a result of the strong cytopathic effect associated with cellular FV Env expression. Finally, in view of the potential use of FV as reagents for gene transfer, determining the cellular receptor(s) of the FV Env is essential. These and other questions are the topic of ongoing research characterizing this unique retroviral glycoprotein.

Acknowledgements. We greatly appreciate the communication of results before publication by A. Saïb and M.L. Linial. Thanks also to H. Zentgraf, T. Wilk, and M. Löchelt for EM pictures. This work was supported by DFG (Li621/2–1, Li621/2–3, and Europäisches Graduiertenkolleg "Gene regulation in and by microbial pathogens") to D.L. and P.A.G. Studies on foamy viruses were supported by a developmental grant from the National Institutes of Health (NIH-K08-AI-01380).

References

Baldwin DN, Linial ML (1998) The roles of Pol and Env in the assembly pathway of human foamy virus. J Virol 72:3658–3665

Bansal A, Shaw KL, Edwards BH, Goepfert PA, Mulligan MJ (2000) Characterization of the R572T point mutant of a putative cleavage site in human foamy virus Env. J Virol 74:2949–2954

Bodem J, Löchelt M, Delius H, Flügel RM (1998) Detection of subgenomic cDNAs and mapping of feline foamy virus mRNAs reveals complex patterns of transcription. Virology 244:417–426

Bullough PA, Hughson FM, Skehel JJ, Wiley DC (1994) Structure of influenza haemagglutinin at the pH of membrane fusion. Nature 371:37–43

Carr CM, Chaudhry C, Kim PS (1997) Influenza hemagglutinin is spring-loaded by a metastable native conformation. Proc Natl Acad Sci USA 94:14306–14313

Carr CM, Kim PS (1993) A spring-loaded mechanism for the conformational change of influenza hemagglutinin. Cell 73:823–832

Chan DC, Fass D, Berger JM, Kim PS (1997) Core structure of gp41 from the HIV envelope glycoprotein. Cell 89:263–273

Deng H, Liu R, Ellmeier W, Choe S, Unutmaz D, Burkhart M, Di Marzio P, Marmon S, Sutton RE, Hill CM, Davis CB, Peiper SC, Schall TJ, Littman DR, Landau NR (1996) Identification of a major co-receptor for primary isolates of HIV-1. Nature 381:661–666

Dragic T, Litwin V, Allaway GP, Martin SR, Huang Y, Nagashima KA, Cayanan C, Maddon PJ, Koup RA, Moore JP, Paxton WA (1996) HIV-1 entry into CD4+ cells is mediated by the chemokine receptor CC-CKR-5. Nature 381:667–673

Epand RM, Epand RF (2001) Factors contributing to the fusogenic potency of foamy virus. Biochem Biophys Res Commun 284:870–874

Fass D, Davey RA, Hamson CA, Kim PS, Cunningham JM, Berger JM (1997) Structure of a murine leukemia virus receptor-binding glycoprotein at 2.0 angstrom resolution. Science 277:1662–1666

Fass D, Harrison SC, Kim PS (1996) Retrovirus envelope domain at 1.7 angstrom resolution. Nat Struct Biol 3:465–469

Feng Y, Broder CC, Kennedy PE, Berger EA (1996) HIV-1 entry cofactor: functional cDNA cloning of a seven-transmembrane, G protein-coupled receptor. Science 272:872–877

Fischer N, Heinkelein M, Lindemann D, Enssle J, Baum C, Werder E, Zentgraf H, Muller JG, Rethwilm A (1998) Foamy virus particle formation. J Virol 72:1610-1615

Flint S, Enquist L, Krug R, Racaniello V, Skalka A (2000) Principles of virology: molecular biology, pathogenesis, and control. American Society for Microbiology, Washington DC.

Flügel RM, Rethwilm A, Maurer B, Darai G (1987) Nucleotide sequence analysis of the env gene and its flanking regions of the human spumaretrovirus reveals two novel genes. EMBO J 6:2077–2084

Giron ML, de The H, Saïb A (1998) An evolutionarily conserved splice generates a secreted env-Bet fusion protein during human foamy virus infection. J Virol 72:4906–4910

Giron ML, Rozain F, Debons Guillemin MC, Canivet M, Peries J, Emanoil Ravier R (1993) Human foamy virus polypeptides: identification of env and bel gene products. J Virol 67:3596–3600

Goepfert PA, Shaw K, Wang G, Bansal A, Edwards BH, Mulligan MJ (1999) An endoplasmic reticulum retrieval signal partitions human foamy virus maturation to intracytoplasmic membranes. J Virol 73:7210–7217

Goepfert PA, Shaw KL, Ritter GD Jr., Mulligan MJ (1997) A sorting motif localizes the foamy virus glycoprotein to the endoplasmic reticulum. J Virol 71:778–784

Goepfert PA, Wang G, Mulligan MJ (1995) Identification of an ER retrieval signal in a retroviral glycoprotein. Cell 82:543–544

Hunter E (1997) Viral entry and receptors. In: Coffin JM, Hughes SH, Varmus HE (eds) Retroviruses. Cold Spring Harbor Laboratory Press, New York

Lindemann D, Bock M, Schweizer M, Rethwilm A (1997) Efficient pseudotyping of murine leukemia virus particles with chimeric human foamy virus envelope proteins. J Virol 71:4815–4820

Lindemann D, Pietschmann T, Picard-Maureau M, Berg A, Heinkelein M, Thurow J, Knaus P, Zentgraf H, Rethwilm A (2001) A particle-associated glycoprotein signal peptide essential for virus maturation and infectivity. J Virol 75:5762–5771

Lindemann D, Rethwilm A (1998) Characterization of a human foamy virus 170-kilodalton Env-Bet fusion protein generated by alternative splicing. J Virol 72:4088–4094

Löchelt M, Flügel RM, Aboud M (1994) The human foamy virus internal promoter directs the expression of the functional Bel 1 transactivator and Bet protein early after infection. J Virol 68:638–645

Mergia A (1994) Simian foamy virus type 1 contains a second promoter located at the 3' end of the env gene. Virology 199:219–222

Netzer KO, Rethwilm A, Maurer B, ter Meulen V (1990) Identification of the major immunogenic structural proteins of human foamy virus. J Gen Virol 71:1237–1241

Peisajovich SG, Shai Y (2002) High similarity between reverse-oriented sequences from HIV and foamy virus envelope glycoproteins. AIDS Res Hum Retroviruses 18:309–312

Pietschmann T, Heinkelein M, Heldmann M, Zentgraf H, Rethwilm A, Lindemann D (1999) Foamy virus capsids require the cognate envelope protein for particle export. J Virol 73:2613–2621

Pietschmann T, Zentgraf H, Rethwilm A, Lindemann D (2000) An evolutionarily conserved positively charged amino acid in the putative membrane-spanning domain of the foamy virus envelope protein controls fusion activity. J Virol 74:4474–4482

Samson M, Libert F, Doranz BJ, Rucker J, Liesnard C, Farber CM, Saragosti S, Lapoumeroulie C, Cognaux J, Forceille C, Muyldermans G, Verhofstede C, Burtonboy G, Georges M, Imai T, Rana S, Yi Y, Smyth RJ, Collman RG, Doms RW, Vassart G, Parmentier M (1996) Resistance to HIV-1 infection in Caucasian individuals bearing mutant alleles of the CCR-5 chemokine receptor gene. Nature 382:722–725

Shaw KL, Lindemann D, Mulligan MJ, Goepfert PA (2003) Foamy virus envelope glycoprotein is sufficient for particle budding and release. J Virol 77: (in press)

Skehel JJ, Wiley DC (2000) Receptor binding and membrane fusion in virus entry: the influenza hemagglutinin. Annu Rev Biochem 69:531–569

Swanstrom R, Wills J (1997) Synthesis, assembly, and processing of viral proteins. In: Coffin JM, Hughes SH and Varmus HE (eds) Retroviruses. Cold Spring Harbor Laboratory Press, New York

Tobaly-Tapiero J, Bittoun P, Neves M, Guillemin MC, Lecellier CH, Puvion-Dutilleul F, Gicquel B, Zientara S, Giron ML, de The H, Saïb A (2000) Isolation and characterization of an equine foamy virus. J Virol 74:4064–4073
Wang G, Mulligan MJ (1999) Comparative sequence analysis and predictions for the envelope glycoproteins of foamy viruses. J Gen Virol 80:245–254
Weissenhorn W, Dessen A, Harrison SC, Skehel JJ, Wiley DC (1997) Atomic structure of the ectodomain from HIV-1 gp41. Nature 387:426–430
Wilk T, de Haas F, Wagner A, Rutten T, Fuller S, Flügel RM, Löchelt M (2000) The intact retroviral Env glycoprotein of human foamy virus is a trimer. J Virol 74:2885–2887
Wilk T, Geiselhart V, Frech M, Fuller SD, Flügel RM, Löchelt M (2001) Specific interaction of a novel foamy virus env leader protein with the n-terminal gag domain. J Virol 75:7995–8007
Zemba M, Wilk T, Rutten T, Wagner A, Flugel RM, Lochelt M (1998) The carboxy-terminal p3 gag domain of the human foamy virus gag precursor is required for efficient virus infectivity. Virology 247:7–13

Chapter 6

Foamy Virus Vectors

A. Mergia, M. Heinkelein

Abstract Gene therapy is a promising novel treatment for a variety of human diseases. Successful application of gene therapy requires the availability of vehicles with the ability to efficiently deliver and express genes. Viral vectors are efficient means of transferring a gene of interest into target cells. Current available vehicles for gene transfer are either inefficient or potentially unsafe for human gene therapy applications.

A. Mergia (✉)
Department of Pathobiology, College of Veterinary Medicine,
University of Florida, Gainesville, FL 32610, USA
E-mail: Mergiaa@mail.vetmed.ufl.edu

M. Heinkelein
Institut für Virologie und Immunbiologie,
Universität Würzburg, Würzburg, Germany

Foamy viruses offer a fresh alternative vector system for gene transfer with the potential to overcome the concerns of the current vectors. Foamy viruses are nonpathogenic and have a broad host range with the ability to infect various types of cells from different species. Foamy virus replication is distinct and may provide an edge for foamy virus vector usage over other retroviral vectors. These features offer the foamy vectors unique opportunities to deliver several genes into a number of different cell types in vivo safely and efficiently. The principal problems for the design of foamy virus vectors have been solved, and several foamy virus vectors that efficiently transduce a variety of cell types are available. This chapter reviews specific features of foamy virus vector systems and recent advances in the development and use of these vectors.

1 Introduction

Successful treatment for diseases such as cancer and AIDS will likely involve particular combinations of drug therapy, immunotherapy, and gene therapy. Gene therapy is a novel approach currently under development for treatment of a variety of diseases (Anderson 1998). This novel therapy offers a fresh alternative, which endows a therapeutic effect by introduction of a particular gene into targeted cells. The effect of gene therapy in the host will depend, in part, on the efficiency of the vector system used to deliver the effector molecule. Numerous vector constructions have been used to deliver and express genes in human cells. Biological vectors for gene therapy are constructed from several viruses that include herpes simplex virus type 1, adenovirus, adenoassociated virus, and retroviruses (Kay et al. 2001; Robbins et al. 1998; Romano et al. 2000; Smith 1999; Wolfe et al. 1999; Wu and Ataai 2000). Among these vectors, retroviruses are most widely used as an efficient means for introducing foreign DNA into the cell genome. The life cycle of retroviruses involves stable integration of viral genetic material into the host genome; expression of viral genes is, in part, regulated by host cell factors (Coffin et al. 1997). These features of retroviruses make them useful for selected applications of gene transfer in animals and humans. Retroviral vectors are significant for basic biological investigations such as studies of cell lineage, mechanisms of carcinogenesis, and developmental and tissue-specific regulation of defined DNA sequences as well as clinical uses for gene therapy (Varmus 1988). Clinical applications of retroviral vectors include recent human studies for the treatment of inherited monogenetic disorders, infectious diseases, and cancer (Anderson 1998; Gunji et al. 2000; Morgan and Blaese 2000; Mountain 2000; Rochlitz 2001).

Vectors have been derived from murine retroviruses [e.g., murine leukemia virus (MuLV)] and from avian retroviruses [e.g., reticuloendothelial virus (REV) and avian sarcoma-leukemia virus (ASLV)] (Eglitis and Anderson 1988). To broaden the cell tropism of MuLV vectors, molecular clones of amphotropic MuLV vectors have been engineered to transduce heterologous genes into cultured cells and into several mammalian species including primates (Miller and Buttimore 1986). Furthermore, higher-titer vectors with a broad host range can be obtained by pseudotyping with the vesicular stomatitis virus G (VSV-G) envelope protein (Yee et al. 1994). However, it appears that, as a result of the limited infection of primate primary cells by MuLV, the efficiency of gene transfer in primates based on amphotropic MuLV vectors is low (Kantoff et al. 1988; Miller 1992). This may be attributed, in part, to the fact that gammaretrovirus replication requires at least one round of cell division for proviral integration into the target cell (Miller et al. 1990; Roe et al. 1993; Springett et al. 1989). Thus these vectors cannot be used for gene transfer to nondividing or growth-arrested cells, severely restricting their potential usefulness. In addition, the duration of gene expression of the gene of interest in vivo with these vectors is often significantly short.

In contrast to the simpler gammaretroviruses, nonproliferating cells such as terminally differentiated macrophages, mucosal dendritic cells, and T lymphocytes are targets of lentivirus infection (Fauci 1996; Levy 1993). These viruses infect nondividing cells because the preintegration complex is recognized by the cell nuclear import machinery and actively transported through the nuclear pore, resulting in integration of proviral DNA, which occurs without the need for cell division (Bukrinsky et al. 1992; Gallay et al. 1996; Lewis and Emerman 1994), whereas MuLV and vectors developed from it cannot transverse an intact nuclear envelope, precluding integration in the absence of mitosis (Lewis and Emerman 1994; Roe et al. 1993). Human immunodeficiency virus (HIV)-based vector systems exploiting this property have been shown to transfer genes to nondividing cells and terminally differentiated cells in vitro, which allow for the efficient in vivo delivery, integration, and long-term expression of transgenes into nonmitotic targets such as adult neurons (Naldini et al. 1996a,b). However, because of the potential risk of generating replication-competent virus, the usage of a HIV vector for gene therapy would remain a public health issue. The animal lentivirus vectors can be pseudotyped with VSV-G envelope to deliver genes into human cells (Federico 1999; Klimatcheva et al. 1999; Poeschla et al. 1998; Schnell et al. 2000). Nevertheless, concern about the safety and efficacy of gene therapy with animal lentiviruses needs to be addressed.

As an alternative to lentivirus vectors, foamy viruses (FVs) may provide a safe and efficient means of targeting various cell types, satisfying current concerns with existing retrovirus vectors. FVs have several inherent features that make them ideal for constructing retroviral vectors to deliver heterologous genes in many species and several cell types. These viruses can be propagated efficiently in various cell types of several species (Hill et al. 1999; Mergia et al. 1996). Cultured epithelial and fibroblast cells as well as lymphoid and neurally originated cells support the growth of FVs. Each FV isolate also infects several mammalian species, and in infected animals virus has been recovered from many organs (Hooks and Detrick-Hooks 1981; von Laer et al. 1996; Schmidt et al. 1997). To date, no disease has been connected to FV in naturally as well as experimentally infected animals (Weiss 1988; Flügel 1991; Linial 2000; Mergia and Luciw 1991; Schmidt et al. 1997). Humans are not a natural host of FV infection; however, occupational infections with primate FV isolates have been reported (Schweizer et al. 1995, 1997; Callahan et al. 1999; Heneine et al. 1998; Sandstrom et al. 2000). Animal caretakers accidentally infected with simian FV remain healthy and no secondary human-to-human transmission could be detected (Heneine et al. 1998). The reader is also referred to the chapter by Heneine in this volume for a detailed discussion of FV transmission. The size of the genome of FV (13 kb) implies that more heterologous DNA can be accommodated in FV-based vectors than the oncoretrovirus and the lentivirus vectors (8–10 kb). Furthermore, recent data demonstrated that reverse transcription occurs late in the replication cycle of FV before the virion is released from the infected cell and virions contain a large amount of reverse-transcribed DNA (Moebes et al. 1997; Yu et al. 1996, 1999). This may provide an edge for FV vectors over the orthoretroviral vectors in cases in which the barrier for gene transfer is at a point of reverse transcription. Similar to HIV DNA, FV DNA can enter the nucleus of G1/S-phase-arrested cells, indicating that FV vectors may be utilized to deliver genes into nondividing and terminally differentiated cells (Saïb et al. 1997). In support of this notion, FV vectors were shown to transduce nondividing cells more efficiently than the oncoretroviral vectors (Hirata et al. 1996; Mergia et al. 2001).

2
Foamy Virus Replication and Strategies for Vector Construction

FVs have a complex genome that encodes the virion structure genes gag, pol, and env, as well as two large open reading frames (ORFs) located at the 3' end of env (Herchenröder et al. 1994; Kupiec et al. 1991; Mergia and Luciw 1991; Renne et al. 1993; Renshaw and Casey 1994b; Tobaly-Tapiero et al. 2000; Winkler et al. 1997). Three ORFs are identified for PFV (formerly designated as human foamy virus) in the corresponding region (Flügel et al. 1987; Maurer et al. 1988). Viral gene expression in FVs is temporally regulated and involves differential gene expression controlled by two promoters. Transcription occurs initially from an internal promoter located at the 3' end of the env gene to express the genes encoded by the ORFs region (Campbell et al. 1994; Löchelt et al. 1993, 1994; Mergia 1994). The first ORF is a transcriptional transactivator (designated tas, for transcriptional transactivator of spumavirus), that strongly augments gene expression directed by viral promoters (Herchenröder et al. 1995; Keller et al. 1991; Mergia et al. 1991; Renne et al. 1993; Renshaw and Casey 1994a; Rethwilm et al. 1991; Venkatesh et al. 1991). After initial expression, Tas feedback on the internal promoter further amplifies the expression of regulatory genes (Löchelt et al. 1993, 1994; Mergia 1994). Subsequently, Tas acts on the long terminal repeat (LTR) to express the virion structural genes. The very low basal levels of expression directed by the viral promoters and the large transactivation in response to Tas indicate that the tas gene is an important determinant of FV replication. For a detailed discussion on the regulation of FV gene expression the reader is referred to the chapter by Löchelt in this book. Mutational analysis in FVs revealed that the tas gene is essential for viral replication (Baunach et al. 1993; Löchelt et al. 1991; Mergia and Wu 1998). The second ORF (orf-2 or bel-2) overlaps the tas gene, and the LTR. A postulated orf-2 (bel-2) protein has been identified as a 44-kDa protein by some investigators (Giron et al. 1993; Löchelt et al. 1991), whereas others have failed to find an orf-2 product (Baunach et al. 1993; He et al. 1993). However, a Bet protein that is a product of a spliced message containing the first 88 amino acids of Tas fused to the last 390 amino acids of Orf-2 was found to be highly expressed in the cytoplasma of

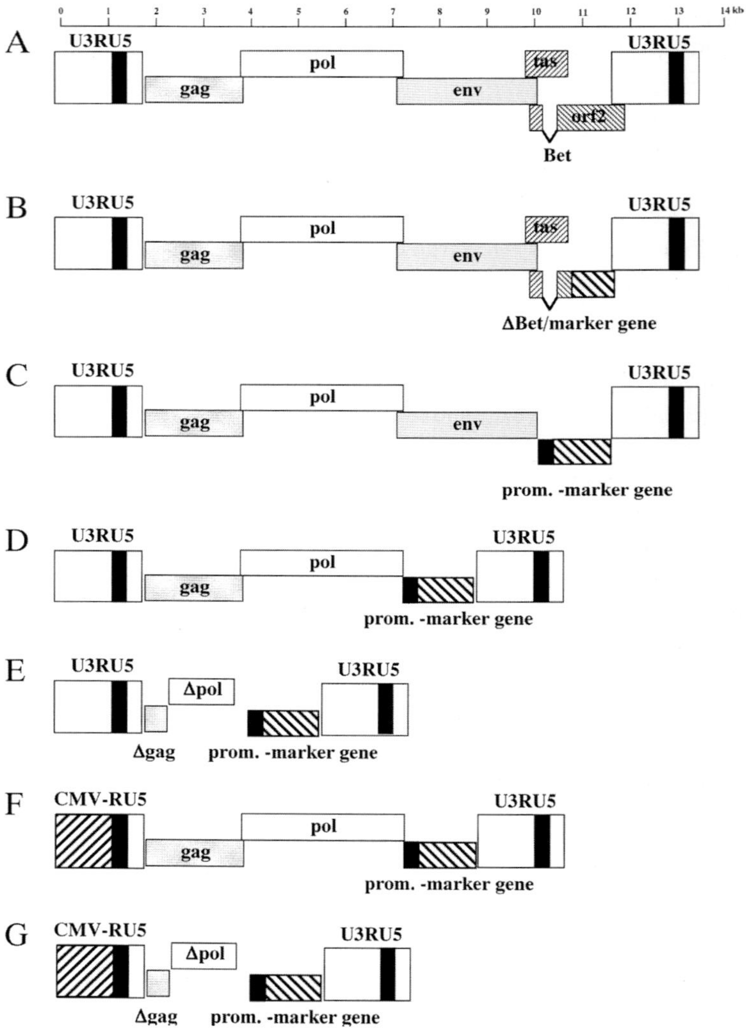

Fig. 1. Schematic representations of published FV vectors. *A*, The provirus genome of wild-type FV. *B*, Replicating FV vectors (Schmidt and Rethwilm 1995). *C*, Replication defective tas-dependent vectors packaged in tas-expressing cell lines (Bieniasz et al. 1997; Mergia and Wu 1998). *D*, Replication defective tas-dependent vectors produced either by cotransfection with replicating helper virus constructs (Russel and Miller 1996; Erlwein et al. 1998; Wu et al. 1998) or by cotransfection with a gag-pol deleted helper virus construct (Russel and Miller 1996). *E*, Replication-defective tas-dependent vectors. In these vectors the coding sequences were removed except the

FV-infected cells (Baunach et al. 1993; Hahn et al. 1994; Löchelt et al. 1991; Muranyi and Flügel 1991). The function of Bet or orf-2 has not been fully elucidated. Using a sensitive assay, Yu and Linial (Yu and Linial 1993) determined that mutations in bel-2 decrease cell-free viral transmission by about 10-fold, suggesting that either Bet or the orf-2 gene product may play a role in efficient cell-free viral transmission similar to Vif of the lentiviruses. Bet may also have a role in persistent infection and/or in preventing productive superinfection (Bock et al. 1998; Callahan et al. 1999; Saïb et al. 1995). Several studies, however, clearly indicate that FVs that are defective for the orf-2 region can replicate in cell culture (Baunach et al. 1993; Mergia and Wu 1998).

Based on the information of the molecular mechanisms of FV replication, vectors that are replication-competent were constructed by replacing a portion of the orf-2 with a reporter gene and showed efficient levels of foreign gene expression (Schmidt and Rethwilm 1995; Fig. 1, B). Several investigators also developed replication-defective vectors by substituting the env and/or the tas-orf-2 region with marker genes (Bieniasz et al. 1997; Erlwein et al. 1998; Heinkelein et al. 1998; Hirata et al. 1996; Mergia and Wu 1998; Russell and Miller 1996; Wu et al. 1998; Fig. 1, C–E). These vectors can be packaged by cotransfection with helper plasmids and are able to transduce a wide variety of cells. Improved FV vectors were designed by replacing the U3 region of the 5' LTR of the virus with the CMV promoter (Fischer et al. 1998; Heinkelein et al. 1998; Trobridge and Russel 1998; Schenk et al. 1999; Wu and Mergia 1999; Fig. 1, F–G). This allows efficient Tas independent transcription of vector genomes, which leads to higher vector production.

regions containing *cis*-acting elements required for FV vectors. These vectors were produced by cotransfection with replicating helper virus constructs (Erlwein et al. 1998; Wu et al. 1998). *F*, Tas-independent replication-defective vectors generated either by cotransfection with a packaging plasmid containing gag/pol/env under the control of CMV promoter (Trobridge and Russel 1998; Wu and Mergia 1999) or by cotransfection with packaging constructs in which gag/pol and env genes are on two separate plasmids (Heinkelein et al. 1998; Wu and Mergia 1999). *G*, Tas-independent replication-defective vector constructs where all the coding sequences for viral genes were deleted (Heinkelein et al. 1998, 2002a; Park et al. 2002; Trobridge et al. 2002b). Vector particles were generated by supplying the gag, pol, and env gene products in trans

3 Foamy Virus Vector Systems

3.1 Replication-Competent Foamy Virus Vectors

Schmidt and Rethwilm constructed several replication-competent FV vectors by replacing part of the ORF-2 (bel-2) of an infectious molecular clone of PFV with marker genes (Schmidt and Rethwilm 1995). In these vectors a region of ORF-2 was deleted immediately downstream the tas gene up to the polypurine tract (PPT) and a small polylinker was introduced to facilitate insertion of foreign genes that are expressed as fusion proteins with a C-terminally truncated Bet-protein. A coding sequence for the self-cleaving foot and mouth disease virus 2A protease was included to allow cleavage of a foreign gene product from the fusion protein. Alternatively, an internal ribosomal entry site was inserted, allowing independent translation and expression of a foreign gene. Furthermore, replication-competent foamy vectors were constructed by placing a marker gene under the control of a foreign promoter (Schmidt and Rethwilm 1995; Meiering et al. 2000). With the replication-competent vectors several indicator genes, including the chloramphenicol acetyltransferase (CAT), luciferase, and gene for green fluorescent protein (GFP) were tested for efficiency of transduction and expression. Infection of cells with these vectors revealed slightly reduced replication kinetics compared to wild-type virus. Titers ranging from 10^4 to 10^7 infectious units per ml can easily be produced with these replication-competent vectors. Infected cells expressed the marker genes efficiently. The stability of the inserted foreign genes within the replicating vector depended not only on the size but apparently also on the primary sequence of the inserted genes. Heterologous DNAs up to 1.2 kb were maintained for at least 10 cell-free passages of the vector, whereas other inserts of about 1 kb were deleted very rapidly (Nestler et al. 1997). Currently, use of the replicating vector system for suicide gene therapy for cancer treatment is under investigation (Nestler et al. 1997). Replication-competent vectors based on a nonpathogenic virus can also be important for the development of effective recombinant vaccines. In addition, these vectors may be of use in studies of tissue distribution of FV in infected animals, perhaps leading to understanding why FVs are nonpathogenic.

3.2
Replication-Defective Vectors

The first replication defective vectors were published by Russell and Miller based on PFV by replacing a portion of the env gene and the internal promoter with reporter genes (Russell and Miller 1996). The foreign genes were placed under the control of heterologous promoters for expression. These vector genomes can be packaged into viral particles by cotransfection with helper plasmids that complement for defects of env and tas gene products. The vector particles had titers ranging from 1×10^3 to 5×10^4/ml and were found able to transduce a wide variety of vertebrate cells by integration of the vector genome. Similar replication-defective FV vectors were also developed based on either the simian FV type 1 (SFV-1) or PFV by removing the entire env-tas and portion of orf-2 region and replacement with marker gene-expressing cassettes (Erlwein et al. 1998; Heinkelein et al. 1998; Wu et al. 1998). Others have constructed foamy vectors by substituting only the accessory genes region with a variety of reporter genes (Bieniasz et al. 1997; Mergia and Wu 1998). Transfection of these vectors into a cell line expressing the transactivator gene can generate vector particles up to 2×10^5/ml by *trans*-complementation.

A tas-independent foamy vector can be developed by exchanging the U3 region of the 5' LTR with a constitutively expressing strong promoter. Traditional vectors were constructed with the 3' and 5' LTRs, the *cis*-acting packaging sequences and the PPT, which is required for the synthesis of the second DNA strand. Recent advances have improved retroviral vectors by replacing the U3 domain of the 5' LTR with the powerful CMV promoter (Soneoka et al. 1995; Naviaux et al. 1996; Kim et al. 1998). The production of high-titer helper-free retroviral vectors can be accomplished with the CMV promoter driving the expression of both the packaging components and the vector genome. This system takes advantage of high-level transfectability and strong E1A-mediated stimulation of CMV-programmed transcription of the human kidney-derived 293 cell line. To improve the efficiency of FV vector transduction, SFV-1 and PFV vectors were designed in which the U3 regions of the 5' LTRs were replaced with CMV promoter (Fischer et al. 1998; Heinkelein et al. 1998; Trobridge and Russel 1998; Schenk et al. 1999; Wu and Mergia 1999). For these vectors the tas gene is not required, because expression of the genome is under the control of the heterologous promoter. On reverse transcription of the genome, the 5' U3 region is restored, generating double-stranded vector DNA that integrates into the host genome. A FV that is devoid of the tas gene is replication defective (Löchelt et al. 1991; Baunach et al. 1993; Mergia and Wu 1998). In the host cell the vector genome is no longer transcribed in the absence of tas and the gene of interest is expressed from a heterologous internal promoter. Therefore, the usage of the CMV promoter that constitutively expresses the genome will avoid the need of the tas gene product for foamy vector construction, restricting viral replication to one cycle. In fact, no wild-type virus was generated with the tas-independent FV vector system as a result of potential recombination between the vector and the packaging DNA (Trobridge and Russel 1998). In addition, internal sequences containing Tas-responsive elements and the TATA box have been deleted from the 3' U3 region of further improved FV vectors (Heinkelein et al. 2002a; Park et al. 2002; Trobridge et al. 2002b). This deletion completely eliminated the risk of activation of the FV U3 promoter in the absence of Tas by cellular transcription factors in the target cell. The lack of virus replication in the absence of tas and the intrinsic nonpathogenic nature of FV argue that these viruses make an ideal, safe vector system for gene therapy.

3.3 Production of High-Titer Virus Stocks

With the tas-independent vector system titers of 5×10^6 vector particles/ml of FV vector production can be achieved in 293T cells (Wu and Mergia 1999). To effectively deliver a gene of interest high-titer vector stocks are required. Vector particles can be concentrated to high titers by methods such as centrifugation or by filtering through size-exclusion pores. The envelope proteins of retroviruses are, in general, fragile and viruses are prone to loss of infectivity when subjected to methods of concentration. However, it is possible to pseudotype retroviral particles that have incorporated the VSV-G envelope protein (Akkina et al. 1996; Bartz and Vodicka 1997; Burns et al. 1993). Retroviral vectors pseudotyped with VSV-G protein possess a wide host range and can be propagated in a wide variety of cell types from different species (Yee et al. 1994). Furthermore, retroviral vector particles containing VSV-G envelopes are less fragile than those containing retroviral envelopes, enhancing pseudotyped vector particle concentration to high titer by centrifugation without loss of infectivity (Yee et al. 1994). Interestingly, attempts to pseudotype FV vectors with VSV-G resulted in a substantial (100-fold) reduction in gene transduction, indicating that VSV-G cannot be used to concentrate FV vector particles (Wu and Mergia 1999). This observation is consistent with the finding that extracellular FV particles can only be detected in the presence of the FV envelope protein (Baldwin and Linial 1998; Fischer et al. 1998). Encapsidation and release of FV particles is, therefore, complex and unique among retroviruses. For a discussion on the peculiarities of FV particle release and Gag-Env interaction see the chapter by Lindemann and Goepfert in this volume. Fortunately for FV vector development, these viruses can be concentrated by centrifugal ultrafiltration or by centrifugation without loss of infectivity (Hill et al. 1999; Trobridge et al. 2002b).

4 *Cis*-Acting Sequences

4.1 *Cis*-Acting Sequences Required to Transfer Foamy Virus Vectors

Although a major breakthrough in the characterization of the molecular mechanism of FV replication has been made (see the chapter by Rethwilm, this volume), important features necessary for FV vector development have just begun to be uncovered. The principal packaging signal of MuLV, the major retroviral vector used in gene delivery, is located in the 5' leader sequence of the viral RNA between the major splice donor site and the gag ATG (Mann et al. 1983). MuLV vectors containing the 5' leader sequence are capable of generating viral particles of significant titer (Bender et al. 1987). The efficiency of packaging is further enhanced by at least one order of magnitude when 5' gag sequences are present in the vector genome (Armentano et al. 1997; Bender et al. 1987), although a more recent study indicated that the presence of the major 5' splice donor may be sufficient (Hildinger et al. 1999). Similarly, the 5' untranslated region of FV is important for FV vector construction. Introduction of deletions in the leader region between the 5' LTR and the start of the gag gene are deleterious for virus replication and completely abrogate gene transfer by FV vectors (Erlwein et al. 1998; Heinkelein et al. 1998; Wu et al. 1998). With strategies of vector development that had been applied to orthoretroviruses, FV vectors have been constructed containing se-

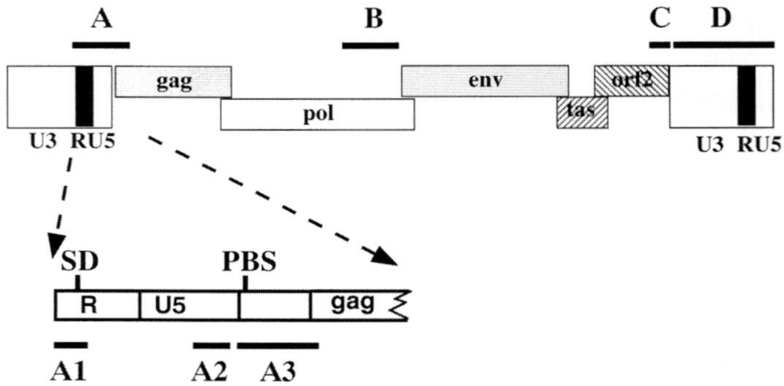

Fig. 2. *Cis*-acting sequences required for FV vectors. *A*, *Cis*-acting elements located at the 5' end of the genome. In addition to the *cis*-acting elements required for transcription, reverse transcription, and integration, sequences important for FV replication and vector construction are identified within this region: *A1*, Element in the R region that is required for Gag and Pol expression. *A2*, *Cis*-element in the U5 region, which is required for proteolytic processing of the capsid. *A3*, RNA dimerization sequences. *B*, *Cis*-acting sequence located in the pol gene, which is critical for FV vector transfer. *C*, PPT. *D*, 3' LTR. *PBS*, primer binding site; *SD*, major 5' splice donor

quences of the 5' and 3' LTRs, the PPT, and the 5' leader region extending into the 5' end of gag sequence (Erlwein et al. 1998; Heinkelein et al. 1998; Wu et al. 1998). However, these vectors failed to transduce marker genes, implicating that additional *cis*-acting sequences were required. These observations led to the discovery of a second *cis*-acting element located in the pol gene that is critical for vector construction (Erlwein et al. 1998; Heinkelein et al. 1998; Wu et al. 1998; Fig. 2).

4.2 *Cis*-Acting Sequences in the pol Gene

For SFV-1, the *cis*-acting element that is critical for vector construction is located at the 3' end of the pol gene between positions 5,185 and 5,674 of the RNA (pre-) genome (Wu et al. 1998). In PFV, the *cis*-acting element has been mapped between positions 4,861 and 5,540, overlapping the SFV-1 *cis*-acting region by 355 nucleotides (Erlwein et al. 1998). Others have located the *cis*-acting element to an approximately 2-kb sequence in the 3' pol region of PFV (Heinkelein et al. 1998,2002a). This 2-kb sequence encompasses the shorter *cis*-acting regions mentioned above. Comparative analysis of the pol genes of primate FVs shows a

85% identity on the nucleotide level including the 3' end of pol, which contains the *cis*-acting element required for vector construction (Mergia and Luciw 1991). Analysis of the RNA for the *cis*-acting region of SFV-1 predicts a stable secondary structure implicating potential interaction sites for viral or cellular proteins (A. Mergia, unpublished observation). Further analysis is necessary to determine the exact sequence(s) of the *cis*-acting element in the pol gene and its precise function in vector transfer.

The necessity of sequences in the pol gene for the transfer of FV vectors has not been found for any other retrovirus. The exact function of the *cis*-acting element at the 3' end of pol in FV replication is not known. It has been shown that the element in question serves no role in either gene expression or posttranscriptional regulation (Heinkelein et al. 1998). Interestingly, a second PPT, which is most likely used as a second site of initiation of plus-strand DNA synthesis, is located within the *cis*-acting region of SFV-1 and the 2-kb region of PFV (Neumann-Haefelin et al. 1986; Tobaly-Tapiero et al. 1991). This PPT is conserved in all FVs including the bovine and feline isolates. For HIV, the central PPT is required for optimal replication (Charneau et al. 1992; Zennou et al. 2000). It is not known whether the second PPT is necessary for FV replication to the extent that it has to be included in vector construction for efficient gene delivery. However, gene transduction can be accomplished with FV vectors in the absence of the pol gene PPT, suggesting that the *cis*-acting element required for vector construction may have a different function (Erlwein et al. 1998; Heinkelein et al. 1998). There are two important observations that may help define the role of the *cis*-acting element of pol. First, RNAse protection experiments revealed that the vector RNA genome is protected fivefold less in the construct that lacks the *cis*-acting element (Heinkelein et al. 1998). Second, protein analysis of cell free vector particles devoid of the *cis*-acting element showed no cleavage of the pr71 Gag precursor (Heinkelein et al. 1998). The cleavage of pr71 Gag by the pol encoded FV protease is critical for FV infectivity (Enssle et al. 1997; Konvalinka et al. 1995). In FV the pol gene is expressed independent of gag from a spliced messenger RNA. Recently it was demonstrated that pregenomic RNA is required for incorporation of the Pol protein into the FV particle, suggesting that virion RNA may be a bridging molecule between Gag and Pol during the assembly of the capsid (Heinkelein et al. 2002b). Together these data suggest that the *cis*-acting sequences located in the pol gene can be important either for packaging and/or stabilization of the RNA in the virus particle or for the interaction of the pregenomic RNA with the Pol protein.

4.3 *Cis*-Acting Elements Located at the 5' End of the Genome

The sequences within the 5' untranslated region of retroviral genomes contain important *cis*-elements for many steps in viral replication. This region is complex with sequences that adopt specific secondary structures. Sequences throughout the 5' untranslated region have been implicated in viral RNA encapsidation, diploid genome dimerization, and efficient gag translation (Berkowitz et al. 1996; Coffin et al. 1997; Linial and Miller 1990; Miele et al. 1996). The 5' untranslated region of FVs is highly conserved among primate viruses and similar to orthoretroviruses, the sequence predicts an extensive RNA secondary structure (Mergia et al. 1992; Park and Mergia 2000; Renne et al. 1993). Recently, sequences within the 5' R and U5 region required for gag expression and Gag protein processing within the virion have been described (Heinkelein et al. 2000a; Russell et al. 2001). A region that facilitates FV RNA genome dimerization is located between the primer binding site (PBS) and the start of the gag gene (Erlwein et al. 1997). This region contains a palindromic sequence (UCCCUAGGGA) that is conserved in all primate FVs and critical for the dimerization of the diploid genome (Cain et al. 2001). For orthoretroviruses, the leader region located between the PBS and the start of the gag gene is also important for genome packaging (Berkowitz et al. 1996; Coffin et al. 1997; Linial and Miller 1990). In HIV-1, the major *cis*-acting packaging site is identified in the leader region between the splice donor and the gag initiation codon, and for efficient encapsidation of the genome, sequences upstream of the splice donor site are also required (Aldovini and Young 1990; Clavel and Orenstein 1990; Lever et al. 1989; Kim and O'Rear 1994; Paillart et al. 1996). In contrast, FV (pre-)genomes can efficiently be packaged in the absence of the sequence between the PBS and the gag gene, suggesting that the major packaging element is located elsewhere in the genome (Heinkelein et al. 2000a; Park and Mergia 2000). For FVs, therefore, the *cis*-acting elements for (pre-)genome dimerization and packaging may be localized in separate regions. Similarly, for HIV-1 RNA dimerization is not necessary for genome packaging (Clever and Parslow 1997). The R region of the FV LTR is approximately 170 nucleotides long, and the major splice donor site within R is located approximately 50 bases downstream from the transcription initiation site (Muranyi and Flügel 1991). FV genomes lacking sequences from nucleotide 152 downstream the transcription start to the beginning of the gag gene are packaged efficiently (Heinkelein et al. 2000a), whereas the deletion of the complete 5' untranslated region resulted in a significant reduction in packaging of FV RNA (Heinkelein et al. 2002a). This suggests that one packaging signal sequence may be located upstream from nucleotide position 152 and probably additional sequences elsewhere in the genome. In HIV-2, the sequence between the splice donor site and gag plays a lesser role in genome encapsidation (McCann and Lever 1997). The major *cis*-acting packaging element is found to be immediately upstream to the 5' splice donor site. The encapsidation site for the avian leukosis virus (ALV) also lies upstream of the splice donor site (Berkowitz et al. 1996; Katz et al. 1986). However, in contrast to FVs, the *cis*-acting elements for these viruses are located in the leader region between the primer binding site and the start of the gag gene. The finding that the region between the PBS and the gag gene is not required for RNA genome packaging strengthens the position that FVs are unique and complex with a distinct replication pathway, including virus (pre-)genome encapsidation. Furthermore, it remains to be determined whether the sequences at the 3' end of the pol gene are involved in FV genome packaging.

The packaging of the (pre-)genomic RNA and the incorporation of the Pol protein into the FV capsid have not yet been elucidated in detail (see the chapter by Linial and Eastman, this volume, for a detailed discussion). Hence, the sequences required to transfer FV vectors can only be determined empirically, and specific functions cannot be assigned. Thus far, studies are focused on the RNA (pre-)genome to identify packaging signal sequences. FV replication, however, is distinct from orthoretroviruses in that reverse transcription occurs late in virus replication (Moebes et al. 1997; Yu et al. 1996, 1999). The process of reverse transcription before particle release from the cell appears to be strictly particle associated, and no reverse transcript can be detected in the absence of Gag protein, suggesting that no viral DNA is packaged independently from the preceding packaging of pregenomic RNA (Enssle et al. 1997). However, specific *cis*-acting elements on the viral DNA may be required for FV vector transfer. The presence of these sequences may therefore complicate the identification of genuine RNA packaging signals.

The primary packaging signal sequence for FV is yet to be identified. It is clear, however, that the 5' untranslated region and the *cis*-acting element at the 3' of the pol gene are important for FV vector construction.

5
Packaging Cell Lines

The availability of stable packaging cell lines would allow a scaled-up production of vector stocks for gene therapy. Several packaging cell lines have been established for retroviral vectors. Generating packaging cell lines for vectors based on viruses that are highly cytopathic may be a difficult task to accomplish. Constitutive expression of the cytopathic retrovirus envelope protein is toxic to cells (Lifson et al. 1986; Sodroski et al. 1986). The protease of HIV has also been implicated in toxic effects of infected cells (Kaplan and Swanstrom 1991; Kräusslich et al. 1993). Inducible HIV packaging cell lines have been created to allow gene expression only when needed and thereby limiting the accumulation of toxic viral proteins (Yu et al. 1996). FV infections result in marked cytopathology in a variety of cultured cell lines (Mergia et al. 1996; Mikovits et al. 1996). The cytopathic effect of FV infection includes cell fusion, potentially mediated by the interaction of the *env* gene product and the cellular receptor(s). Establishing a packaging cell line containing the structural genes of a FV can therefore be an inherent stumbling block for FV vector development. Furthermore, the FV vector genome may intracellularly retrotranspose, which could eventually lead to a substantial increase in proviruses (Heinkelein et al. 2000b). This feature of primate FVs can add to the cytopathology (Heinkelein et al. 2000b).

Despite these potential inherent problems, attempts have been made to establish packaging cell lines from parental 293 cells (Wu and Mergia 1999). However, expression of Gag protein has not been demonstrated in these cell lines and an efficient transfer of vector may require the presence of a functional gag gene on the vector genome itself. This may illustrate the difficulties in establishing FV packaging cells lines, which have also been experienced by others (D. Neumann-Haefelin, M. Schweizer, M. McClure, and O. Erlwein, personal communications).

Table 1. Host range of FV and cell types susceptible to FV infection in vitro

Species	Cell line	Cell type	Reference
Human	Primary	Fibroblasts	Hirata et al. 1996; Russel and Miller 1996
	Primary	CD34^{+} cells	Hirata et al. 1996; Trobridge et al. 2002; Zucali et al. 2002
	Primary	Bone marrow stroma cells	Hirata et al. 1996
	Primary	Peripheral lymphocytes	Mergia et al. 2001
	HT1080	Fibrosarcoma	Lindemann et al. 1997; Hill et al. 1999
	MRC-5, KMST-6	Fibroblasts	Schmidt et al. 1995; Mergia et al. 1996
	293, HeLa	Epithelial	Mergia et al. 1996; Hill et al. 1999
	EBV transf.	B-cell blasts	Heinkelein et al., unpublished data
	CEM, HUT78	T cell	Mikovits et al. 1996; Mergia et al. 1996
	Jurkat, A301, H9	T cell	Adachi et al. 1995; Yu et al. 1996
	H92.1.7	Erythroblastoid	Mikovits et al. 1996; Yu et al. 1996
	U937, GCT	Myeloid monocytic	Adachi et al. 1995; Yu et al. 1996; Mergia et al. 1996
	THP-1, KG1, K-562	Myeloid	Adachi et al. 1995; Hirata et al 1996; Mergia et al. 2001; Mikovits et al. 1996
	RD	Rhabdomyosarcoma	Hill et al. 1999
	U87, SK-N-MC, NB-1, IMR32, T98G, A172, U251-MG	Glioblastoma, neuroblastoma	Rethwilm et al. 1990.; Adachi et al. 1995; Mergia et al. 1996; Nestler et al. 1997
	NT2 N	Postmitotic neuron cells	Mergia et al. 2001
	HepG2, Hep3b	Liver epithelial	Nestler et al. 1997; Heinkelein et al., unpublished data
SW480	Colon carcinoma	Adachi et al. 1995	
Monkey	Primary	CD34^{+} cells	Hirata et al. 1996
	COS-7, CV-1, RHFL	Fibroblast	Russell and Miller 1996; Mergia et al. 1996
Vero	Kidney epithelial	Russell and Miller 1996; Mergia et al. 1996	

Table 1. (continued)

Species	Cell line	Cell type	Reference
Mouse	Primary	Bone marrow cells	Hirata et al. 1996; Vassilopoulos et al. 2001
	NIH3T3, L929	Fibroblast	Mergia et al. 1996; Hill et al. 1999
	WOP, LMtk-	Fibroblast	Hill et al. 1999
MEL	Erythroleukemia	Hirata et al. 1996	
Rat	NRK	Kidney epithelial	Russel and Miller 1996
	208F		Russel and Miller 1996
Rat-1	Fibroblast	Hill et al. 1999	
Hamster	BHK-21	Kidney fibroblast	Russel and Miller 1996; Lindemann et al. 1997
CHO-K1	Ovary epithelial	Russel and Miller 1996; Hill et al. 1999	
Mink	Mv.1.Lu	Lung epithelial	Lindemann et al. 1997; Hill et al. 1999
Dog	Cf2Th	Thymus fibroblast	Russel and Miller 1996; Mergia et al. 1996
Cat	CCC-81, AH927		Russell and Miller 1996
CRFK	Fibroblast	Mergia et al. 1996	
Bovine	MDBK	Kidney epithelial	Russell and Miller 1996
Pig	PAE	Aortic epithelial	Hill et al. 1999
PK15	Kidney epithelial	Hill et al. 1999	
Guinea pig	MPK	Kidney	Hill et al. 1999
Dolphin	Sp1 K	Kidney epithelial	Hill et al. 1999
Avian	Primary	Embryo, fibroblasts	Russell and Miller 1996
	QT-6	Quail fibrosarcoma	Mergia et al 1996; Lindemann et al. 1997
CBGQ	Goose fibroblast	Hill et al. 1999	
Reptilian	IgH2	Iguana epithelium	Hill et al. 1999

6 Cell Tropism and Infection of Nondividing Cells

The surface receptor molecule that enables FV to enter cells is not known. As shown in Table 1 FVs have a wide host range and can infect cells of many different lineages (Hill et al. 1999; Mergia et al. 1996). In infected animals, virus appears to be present in several tissues, including brain and peripheral blood mononuclear cells (Hooks and Detrick-Hooks 1981; Schmidt et al. 1997; von-Laer et al. 1996). These observations indicate that the surface receptor is a molecule that is common to many if not all mammalian cells and that virtually any type of cell can be targeted with a FV vector to deliver a gene of interest. Although FVs have a broad host range, viral infection studies in cells arrested at G1/S phase suggested that FV replication is cell cycle dependent (Bieniasz et al. 1995). Contrary to this result, others have reported that gene transduction with a PFV vector in stationary phase cell culture is more efficient than with a MuLV vector (Hirata et al. 1996; Russell and Miller 1996). Furthermore, Saïb et al. were able to demonstrate that FV DNA can enter the nucleus of G1/S phase -arrested cells (Saïb et al. 1997). However, although DNA can enter the nucleus of G1/S phase-arrested cells, no viral gene expression could be observed, suggesting that FV promoters may require cycling cells for functionality. These contradictory observations were recently addressed by Mergia et al. by demonstrating that a high-titer SFV-1-based vector is able to mediate efficient transduction of various arrested and nondividing cell types, including lymphocytes and neurons (Mergia et al. 2001). The reasons for the efficient transduction of nondividing cells in comparison to previous studies with PFV may in part be the usage of a 100-fold higher titer vector and the use of the CMV promoter for expression of the transgene. The CMV promoter was shown previously to be active in a variety of growth-arrested cells. Therefore, FV vectors may be able to provide an efficient method for clinical gene transfer into primary cells. However, further studies are needed to analyze the requirements for efficient transduction of nondividing cells by FV vectors. Recently Vassilopoulos et al. demonstrated efficient PFV vector transduction of human $CD34^+$ hematopoietic progenitors and murine hematopoietic stem cells (Vassilopoulos et al. 2001; Trobridge et al. 2002b). Efficient long-term expression of the vector-encoded transgene in all major hematopoietic cell lineages could be detected in mice that had received transduced murine bone marrow cells. Secondary transplants and integration site analysis confirmed that gene transfer occurred at the stem cell level. Currently the transduction of human SCID repopulating cells (SRC) is under investigation. In these experiments FV vectors were able to efficiently transduce human hematopoietic cells capable of repopulating NOD/SCID mice including cells with pluripotent capacity (Josephson et al. 2002; Zucali et al. 2002; Leurs et al. 2002). Only short time ex vivo manipulation was required for efficient transduction (Josephson et al. 2002; Leurs et al. 2002). Transduction efficacies of human SRCs by FV vectors were comparable to those achieved in parallel for a VSV-G-pseudotyped HIV-derived vector and superior to both HIV as well as MLV vectors pseudotyped with a chimeric GALV/Amphotropic MLV envelope (Leurs et al. 2002). These results establish the utility of FV vectors in clinical important cell types for gene therapy.

7
Conclusions and Future Implications

Gene therapy is an emerging field, and recently published results demonstrating successful gene therapeutic approaches for the treatment of human diseases are very encouraging (Cavazzana-Calvo et al. 2000; Civin 2000; Kouraklis 2000). Although various vector systems are available, there still is a growing demand for suitable gene transfer vehicles. One of the major problems in gene therapy is the low transduction efficiency in target tissues. Current available retroviral vectors are either inefficient or potentially unsafe for human gene therapy applications. FVs offer a fresh alternative vector system for gene transfer. Principle problems for the design of FV vectors have been solved, and several FV vector systems that efficiently transduce a variety of cell types from different species are available. The process of vector development helped to acquire substantial information about FVs, and the vectors are now widely used tools to unravel several unexplained phenomena of the FV replication cycle. Complete understanding of virus replication, particularly (pre-)genome packaging and particle formation, will help to develop efficient FV vectors with minimal genome sequences that allow large-size foreign DNA inserts. Information available thus far indicates that a FV vector can allow foreign DNA inserts of 9 kb at a minimum (Park et al. 2002; Trobridge et al. 2002). Therefore, such a vector can accommodate average-size cDNAs of more than one gene. Furthermore, because humans do not harbor FVs, the potential risk of vector clearance by a preexisting immune response is unlikely. In accidentally infected humans FV persists in primary peripheral blood lymphocytes for at least 20 years, suggesting that FV vector is capable of long-term gene transfer in vivo (Heneine et al. 1998; Schweizer et al. 1997). The duration of expression of genes transduced by FV vectors in vivo has to be explored over a long period. The observation that pathology is not associated with FV infection and the fact that the virus has a broad host range that may include pluripotent human hematopoietic progenitor cells and nondividing cells provide the FVs unique opportunities to deliver genes into a number of different cell types in vivo safely and efficiently.

References

Adachi A, Sakai H, Tokunaga K, Kawamura M (1995) Functional analysis of human spuma retrovirus genome. Virus Genes 11:15–20

Akkina RK, Walton RM, Chen ML, Li QX, Planelles V, Chen IS (1996) High-efficiency gene transfer into CD34+ cells with a human immunodeficiency virus type 1-based retroviral vector pseudotyped with vesicular stomatitis virus envelope glycoprotein G. J Virol 70:2581–2585

Aldovini A, Young RA (1990) Mutations of RNA and protein sequences involved in human immunodeficiency virus type 1 packaging result in production of noninfectious virus. J Virol 64: 1920–1926

Anderson WF (1998) Human gene therapy. Nature 392:25–30

Armentano D, Yu SF, Kantoff PW, von-Ruden T, Anderson WF, Gilboa E (1997) Effect of internal viral sequences on the utility of retroviral vectors. J Virol 61:1647–1650

Bartz SR, Vodicka MA (1997) Production of high-titer human immunodeficiency virus type 1 pseudotyped with vesicular stomatitis virus glycoprotein. Methods 12:337–342

Baldwin DN, Linial ML (1998) The roles of Pol and Env in the assembly pathway of human foamy virus. J Virol 72:3658–3665

Baunach G, Maurer B, Hahn H, Kranz M, Rethwilm A (1993) Functional analysis of human foamy virus accessory reading frames. J Virol 67:5411–5418

Bender MA, Palmer TD, Gelinas RE, Miller AD (1987) Evidence that the packaging signal of Moloney murine leukemia virus extends into the gag region. J Virol 62:1639–1646

Berkowitz RD, Fisher J, Goff SP (1996) RNA packaging. Curr Top Microbiol Immunol 214:177–218

Bieniasz PD, Erlwein O, Aguzzi A, Rethwilm A, McClure MO (1997) Gene transfer using replication-defective human foamy virus vectors. Virology 235:65–72

Bieniasz PD, Weiss RA, McClure MO (1995) Cell cycle dependence of foamy retrovirus infection. J Virol 69:7295–7299

Bock M, Heinkelein M, Lindemann D, Rethwilm A (1998) Cells expressing the human foamy virus (HFV) accessory Bet protein are resistant to productive HFV superinfection. Virology 250:194–204

Bukrinsky MI, Sharova N, Dempsey MP, Stanwick TL, Bukrinskaya AG, Haggerty S, Stevenson M (1992) Active nuclear import of human immunodeficiency virus type 1 preintegration complexes. Proc Natl Acad Sci USA 89:6580–6584

Burns JC, Friedmann T, Driever W, Burrascano M, Yee JK (1993) Vesicular stomatitis virus G glycoprotein pseudotyped retroviral vectors: concentration to very high titer and efficient gene transfer into mammalian and nonmammalian cells. Proc Natl Acad Sci USA 90:8033–8037

Cain D, Erlwein O, Grigg A, Russell RA and McClure MO (2001) Palindromic sequence plays a critical role in human foamy virus dimerization. J Virol 75:3731–3739

Callahan ME, Switzer WM, Matthews AL, Roberts BD, Heneine W, Folks TM, Sandstrom PA (1999) Persistent zoonotic infection of a human with simian foamy virus in the absence of an intact orf-2 accessory gene. J Virol 73:9619–9624

Campbell M, Renshaw-Gegg L, Renne R, Luciw PA (1994) Characterization of the internal promoter of simian foamy viruses. J Virol 68:4811–4820

Cavazzana-Calvo M, Hacein-Bey S, de-Saint-Basile G, Gross F, Yvon E, Nusbaum P, Selz F, Hue C, Certain S, Casanova JL, Bousso P, Deist FL, Fischer A (2000) Gene therapy of human severe combined immunodeficiency (SCID)-X1 disease. Science 288:669–72

Charneau P, Alizon M, Clavel F (1992) A second origin of DNA plus-strand synthesis is required for optimal human immunodeficiency virus replication. J Virol 66:2814–2820

Civin C (2000) Gene therapy in clinical applications. Stem Cells 180:150–156

Clavel F, Orenstein JM (1990) A mutant of human immunodeficiency virus with reduced RNA packaging and abnormal particle morphology. J Virol 64:5230–5234

Clever JL, Parslow TG (1997) Mutant human immunodeficiency virus type 1 genomes with defects in RNA dimerization or encapsidation. J Virol 71:3407–3414

Coffin JM, Hughes SH, Varmus HE, eds (1997) Retroviruses. Cold Spring Harbor Laboratory Press, New York

Eglitis MA, Anderson WF (1988) Retroviral vectors for introduction of genes into mammalian cells. BioTechniques 6:608–614

Enssle J, Fischer N, Moebes A, Mauer B, Smola U, Rethwilm A (1997) Carboxy-terminal cleavage of the human foamy virus gag precursor molecule is an essential step in the viral life cycle. J Virol 71:7312–7317

Erlwein O, Bieniasz PD, McClure MO (1998) Sequences in pol are required for transfer of human foamy virus-based vectors. J Virol 72:5510–5516

Erlwein O, Cain D, Fischer N, Rethwilm A, McClure MO (1997) Identification of sites that act together to direct dimerization of human foamy virus RNA in vitro. Virology 229:251–258

Fauci AS (1996) Host factors and the pathogenesis of HIV-induced disease. Nature 384:529–534

Federico M (1999) Lentiviruses as gene delivery vectors. Curr Opin Biotechnol 10:448–453

Fischer N, Heinkelein M, Lindemann D, Enssle J, Baum C, Werder E, Zentgraf H, Muller JG, Rethwilm A (1998) Foamy virus particle formation. J Virol 72:1610–1615

Flügel RM (1991) Spumaviruses: a group of complex retroviruses. J AIDS 4:739–750

Flügel RM, Rethwilm A, Maurer B, Darai G (1987) Nucleotide sequence analysis of the env gene and its flanking regions of the human spumaretrovirus reveals two novel genes. EMBO J 6:2077–84

Gallay P, Stitt V, Mundy C, Oettinger M, Trono D (1996) Role of the karyopherin pathway in human immunodeficiency virus type 1 nuclear import. J Virol 70:1027–1032

Giron ML, Rozain F, Debons-Guillemin MC, Canivet M, Peries J, Emanoil-Ravier R (1993) Human foamy virus polypeptides: identification of env and bel gene products. J Virol 67:3596–3600

Gunji Y, Ochiai T, Shimada H, Matsubara H (2000) Gene therapy for cancer. Surg Today 30:967–73

Hahn H, Gerald B, Brautigam S, Mergia A, Neumann-Haefelin D, Daniel MD, McClure MO, Rethwilm A (1994) Reactivity of primate sera to foamy virus Gag and Bet proteins. J Gen Virol 75:2635–2644

He F, Sun JD, Garrett ED, Cullen BR (1993) Functional organization of the Bel-1 transactivator of human foamy virus. J Virol 67:1896–1904

Heinkelein M, Schmidt M, Fischer N, Moebes A, Lindemann D, Enssle J, Rethwilm A (1998) Characterization of a cis-acting sequence in the Pol region required to transfer human foamy virus vectors. J Virol 72:6307–6314

Heinkelein M, Thurow J, Dressler M, Imrich H, Neumann-Haefelin D, McClure MO, Rethwilm A (2000a) Complex effects of deletions in the 5' untranslated region of primate foamy virus on viral gene expression and RNA packaging. J Virol 74:3141–3148

Heinkelein M, Pietschmann T, Jarmy G, Dressler M, Imrich H, Thurow J, Lindemann D, Bock M, Moebes A, Roy J, Herchenröder O and Rethwilm A (2000b) Efficient intracellular retrotransposition of an exogenous primate retrovirus genome. EMBO J 19:3436–3445

Heinkelein M, Dressler M, Jarmy G, Rammling M, Imrich H, Thurow J, Lindemann D and Rethwilm A (2002a) Improved primate foamy virus vectors and packaging constructs. J Virol 76:3774–3783

Heinkelein M, Leurs C, Rammling M, Peters K, Hanenberg H and Rethwilm A (2002b) Pregenomic RNA is required for efficient incorporation of Pol polyprotein into foamy virus capsids. J Virol 76:(in press)

Heneine W, Switzer WM, Sandstrom P, Brown J, Vedapuri S, Schable CA, Khan AS, Lerche NW, Schweizer M, Neumann-Haefelin D, Chapman LE, Folks TM (1998) Identification of a human population infected with simian foamy viruses. Nat Med 4:403–407

Herchenröder O, Renne R, Loncar D, Cobb EK, Murthy KK, Schneider J, Mergia A, Luciw PA (1994) Isolation, cloning, and sequencing of simian foamy viruses from chimpanzees (SFVcpz): high homology to human foamy virus (HFV). Virology 201:187–199

Herchenröder O, Turek R, Neumann-Haefelin D, Rethwilm A, Schneider J (1995) Infectious proviral clones of chimpanzee foamy virus (SFVcpz) generated by long

PCR reveal close functional relatedness to human foamy virus. Virology 214:685–689

Hildinger M, Abel KL, Ostertag W and Baum C (1999) Design of 5' untranslated sequences in retroviral vectors developed for medical use. J Virol 73:4083–4089

Hill CL, Bieniasz PD, McClure MO (1999) Properties of human foamy virus relevant to its development as a vector for gene therapy. J Gen Virol 80:2003–2009

Hirata RK, Miller AD, Andrews RG, Russell DW (1996) Transduction of hematopoietic cells by foamy virus vectors. Blood 88:3654–3661

Hoogerbrugge PM, van Beusechem VW, Fischer A, Debree M, Deist Fl, Perignon JL, Morgan G, Gaspar B, Fairbanks LD, Skeoch CH, Moseley A, Harvey M, Levinsky RJ, Valerio D (1996) Bone marrow gene transfer in three patients with adenosine deaminase deficiency. Gene Ther 3:179–83

Hooks JJ, Detrick-Hooks B (1981) Spumavirinae: foamy virus group infections. Comparative aspects and diagnosis. In: Kurstak E, Kurstak C (eds) Comparative Diagnosis of Viral Disease. Academic Press, New York

Josephson NC, Vassilopoulos G, Trobridge GD, Priestley GV, Wood BL, Papayannopoulou T, Russell DW (2002) Transduction of human NOD/SCID-repopulating cells with both lymphoid and myeloid potential by foamy virus vectors. Proc Natl Acad Sci USA 99:8295–8300

Kantoff PW, Freeman SM, Anderson WF (1988) Prospects for gene therapy for immunodeficiency diseases. Annu Rev Immunol 6:581–594

Kaplan AH, Swanstrom R (1991) Human immunodeficiency virus type 1 Gag proteins are processed in two cellular compartments. Proc Natl Acad Sci USA 88:4528–4532

Katz RA, Terry RW, Skalka AM (1986) A conserved cis-acting sequence in the 5' leader of avian sarcoma virus RNA is required for packaging. J Virol 59:163–167

Kay MA, Glorioso JC, Naldini L (2001) Viral vectors for gene therapy: the art of turning infectious agents into vehicles of therapeutics. Nature Med 7:33–40

Keller A, Partin KM, Löchelt M, Bannert H, Flügel RM, Cullen BR (1991) Characterization of the transcriptional trans-activator of human foamy virus. J Virol 65:2589–2594

Kim HJ, Lee K, O'Rear JJ (1994) A short sequence upstream of the 5" major splice site is important for encapsidation of HIV-1 genomic RNA. Virology 198:336–340

Kim VN, Mitrophanous K, Kingsman SM, Kingsman AJ (1998) Minimal requirement for a lentivirus vector based on human immunodeficiency virus type 1. J Virol 72:811–816

Klimatcheva E, Resenblatt JD, Planelles V (1999) Lentiviral vectors and gene therapy. Front Biosci 4:D482–496

Konvalinka J, Löchelt M, Zentgraf H, Flügel RM, Kräusslich HG (1995) Active foamy virus proteinase is essential for virus infectivity but not for formation of a Pol polyprotein. J Virol 69:7264–7268

Kouraklis G (2000) Gene therapy for cancer: from the laboratory to the patient. Dig Dis Sci 45:1045–1052

Kräusslich HG, Ochsenbauer C, Traenckner AM, Mergener K, Facke M, Gelderblom HR, Bosch V (1993) Analysis of protein expression and virus-like particle formation in mammalian cell lines stably expressing HIV-1 gag and env gene products with or without active HIV proteinase. Virology 192:605–617

Kupiec J, Kay A, Hayat M, Ravier R, Peries J, Galibert F (1991) Sequence analysis of the simian foamy virus type 1 genome. Gene 101:185–194
Leurs C, Jansen M, Pollok KE, Heinkelein M, Schmidt M, Wissler M, Lindemann D, von Kalle C, Rethwilm A, Williams DA, Hanenberg H (2002) Comparison of three retroviral vector systems for transduction of NOD/SCID repopulating CD34+ cord blood cells (submitted)
Lever A, Göttlinger H, Haseltine W, Sodroski J (1989) Identification of a sequence required for efficient packaging of human immunodeficiency virus type 1 RNA into virions. J Virol 63:4085–4087
Levy J (1993) Pathogenesis of human immunodeficinecy virus infection. Microbiol Rev 57:183–289
Lewis PF, Emerman M (1994) Passage through mitosis is required for oncoretroviruses but not for the human immunodeficiency virus. J Virol 68:510–516
Lifson JD, Reyes GR, McGrath MS, Stein BS, Engleman EG (1986) AIDS retrovirus induced cytopathology: giant cell formation and involvement of CD4 antigen. Science 232:1123–1127
Linial M (2000) Why aren't foamy viruses pathogenic? Trends Microbiol 8:284–289
Linial ML, Miller AD (1990) Retroviral RNA packaging: sequence requirements and implications. Curr Topics Micro Immunol 157:125–185
Löchelt M, Flügel RM, Aboud M (1994) The human foamy virus internal promoter directs the expression of the functional Bel 1 and Bet protein early after infection. J Virol 68:638–645
Löchelt M, Muranyi W, Flügel RM (1993) Human foamy virus genome posses an internal, Bel-1-dependent and functional promoter. Proc Natl Acad Sci USA 90:7317–7321
Löchelt M, Zentgraf H, Flügel RM (1991) Construction of an infectious DNA clone of the full-length human spumaretrovirus genome and mutagenesis of the bel 1 gene. Virology 184:43–54
Mann R, Mulligan RC, Baltimore D (1983) Construction of a retrovirus packaging mutant and its use to produce helper-free defective retrovirus. Cell 33:153–159
Maurer B, Bannert H, Darai G, Flügel RM (1988) Analysis of the primary structure of the long terminal repeat and the gag and pol genes of the human spumaretrovirus. J Virol 62:1590–1597
McCann EM, Lever AM (1997) Location of cis-acting signals important for RNA encapsidation in the leader sequence of human immunodeficiency virus type 2. J Virol 71:4133–4137
Meiering CD, Comstock KE, Linial ML (2000) Multiple integrations of human foamy virus in persistently infected human erythroleukemia cells. J Virol 74:1718–1726
Mergia A (1994) Simian foamy virus type 1 contains a second promoter located at the 3' end of the env gene. Virology 199:219–222
Mergia A, Chari S, Kolson DL, Goodenow MM, Ciccarone T (2001) The efficiency of simian foamy virus vector type-1 (SFV-1) in nondividing cells and in human PBLs. Virology 280:243–252
Mergia A, Leung NJ, Blackwell J (1996) Cell tropism of the simian foamy virus type 1 (SFV-1). J Med Primatol 25:2–7
Mergia A, Luciw PA (1991) Replication and regulation of primate foamy viruses. Virology 184:475–482

Mergia A, Pratt-Lowe E, Shaw KE, Renshaw-Gegg LW, Luciw PA (1992) cis-Acting regulatory regions in the long terminal repeat of simian foamy virus type 1. J Virol 66:251–257

Mergia A, Shaw KE, Pratt-Lowe E, Barry PA, Luciw PA (1991) Identification of the simian foamy virus transcriptional transactivator gene (taf). J Virol 65:2903–2909

Mergia A, Wu M (1998) Characterization of provirus clones of simian foamy virus type 1 (SFV-1). J Virol 72:817–822

Miele G, Mouland A, Harrison GP, Cohen E, Lever AM (1996) The human immunodeficiency virus type 1 5" packaging signal structure affects translation but does not function as an internal ribosome entry site structure. J Virol 70:944–951

Mikovits JA, Hoffman PM, Rethwilm A, Ruscetti FW (1996) In vitro infection of primary and retrovirus-infected human leukocytes by human foamy virus. J Virol 70:2774–2780

Miller AD (1992) Human gene therapy comes of age. Nature 357:455–460

Miller AD, Buttimore C (1986) Redesign of retrovirus packaging cell lines to avoid recombination leading to helper virus production. Mol Cell Biol 6:2895–2902

Miller DG, Adam MA, Miller AD (1990) Gene transfer by retroviral vectors occurs only in cells that are actively replicating at the time of infection. Mol Cell Biol 10:4239–4242

Moebes A, Enssle J, Bieniasz PD, Heinkelein M, Lindemann D, Bock M, McClure MO, Rethwilm A (1997) Human foamy virus reverse transcription that occurs late in the viral replication cycle. J Virol 71:7305–11

Morgan RA, Blaese RM (1999) Gene therapy: lessons learnt from the past decade. BMJ 319:1310–12

Mountain A (2000) Gene therapy: the first decade. Trends Biotechnol 18:119–128

Muranyi W, Flügel RM (1991) Analysis of splicing patterns of human spumaretrovirus by polymerase chain reaction reveals complex RNA structures. J Virol 65:727–735

Naldini L, Blomer U, Gage FH, Trono D, Verma IM (1996a) Efficient transfer, integration, and sustained long-term expression of the transgene in adult rat brains injected with a lentiviral vector. Proc Natl Acad Sci USA 93:11382–11388

Naldini L, Blomer U, Gallay P, Ory D, Mulligan R, Gage FH, Verma IM, Trono D (1996b) In vivo gene delivery and stable transduction of nondividing cells by a lentiviral vector. Science 272:263–267

Naviaux RK, Costanzi E, Haas M, Verma IM (1996) The pCL vector system: rapid production of helper-free, high-titer, recombinant retroviruses. J Virol 70:5701–5705

Nestler U, Heinkelein M, Lucke M, Meixensberger J, Scheurlen W, Kretschmer A, Rethwilm A (1997) Foamy virus vectors for suicide gene therapy. Gene Ther 4:1270–1277

Neumann-Haefelin D, Schweizer M, Corsten B, Matz B (1986) Detection and characterization of infectious DNA intermediates of a primate foamy virus. J Gen Virol 67:1993–1999

Paillart JC, Berthoux L, Ottmann M, Darlix JL, Marquet R, Ehresmann B, Ehresmann C (1996) A dual role of the putative RNA dimerization initiation site of human immunodeficiency virus type 1 in genomic RNA packaging and proviral DNA synthesis. J Virol 70:8348–8354

Park J, Mergia A (2000) Mutational analysis of the 5' leader region of simian foamy virus type 1. Virology 274:203–212
Park J, Nadeau PE, Mergia A (2002) A minimal genome simian foamy virus type 1 (SFV-1) vector system with efficient gene transfer. Virology 302: 236–244
Poeschla EM, Wong-Staal F, Looney DJ (1998) Efficient transduction of nondividing human cells by feline immunodeficiency virus lentiviral vectors. Nat Med 4:354–357
Renne R, Mergia A, Renshaw-Gegg LW, Neumanm-Haefelin D, Luciw PA (1993) Regulatory elements in the long terminal repeat (LTR) of simian foamy virus type 3. Virol 192:365–369
Renshaw RW, Casey JW (1994a) Analysis of the 5 ' long terminal repeat of bovine syncytial virus. Gene 141:221–224
Renshaw RW, Casey JW (1994b) Transcriptional mapping of the 3' end of the bovine syncytial virus genome. J Virol 68:1021–1028
Rethwilm A, Baunach G, Netzer K-O, Maurer B, Borisch B, ter Maulen V (1990) Infectious DNA of the human spumaretrovirus. Nucleic Acids Res 18:733–738
Rethwilm A, Erlwein O, Baunach G, Maurer B, ter Meulen V (1991) The transcriptional transactivator of human foamy virus maps to the bel 1 genomic region. Proc Natl Acad Sci USA 88:941–945
Robbins P, Tahara H, Ghivizzani S (1998) Viral vectors for gene therapy. Trends Biotechnol 16:35–40
Rochlitz CF (2001) Gene therapy of cancer. Schweiz Med Wochenschr 131:4–9
Roe T, Reynolds TC, Yu G, Brown PO (1993) Integration of murine leukemia virus DNA depends on mitosis. EMBO J 12:2099–2108
Romano G, Micheli P, Pacilio C, Giordano A (2000) Latest developments in gene transfer technology: achievements, perspectives, and controversies over therapeutic applications. Stem Cells 18:19–39
Russell DW, Miller AD (1996) Foamy virus vectors. J Virol 70:217–22
Russell RA, Zeng Y, Erlwein O, Cullen BR, McClure MO (2001) The R region found in the human foamy virus long terminal repeat is critical for both Gag and Pol protein expression. J Virol 75:6817–6824
Saïb A, Koken MH, van der Spek P, Peries J, de The H (1995) Involvement of a spliced and defective human foamy virus in the establishment of chronic infection. J Virol 69:5261–5268
Saïb A, Puvion-Dutilleul F, Schmid M, Peries J, de The H (1997) Nuclear targeting of incoming human foamy virus Gag proteins involves a centriolar step. J Virol 71:1155–61
Sandstrom PA, Phan KO, Switzer WM, Fredeking T, Chapman L, Heneine W, Folks TM (2000) Simian foamy virus infection among zoo keepers. Lancet 355:551–552
Schenk T, Enssle J, Fischer N, Rethwilm A (1999) Replication of a foamy virus mutant with a constitutively active U3 promoter and deleted accessory genes. J Gen Virol 80:1591–1598
Schmidt M, Niewiesk S, Heeney J, Aguzzi A, Rethwilm A (1997) Mouse model to study the replication of primate foamy viruses. J Gen Virol 78:1929–1933
Schmidt M, Rethwilm A (1995) Replicating foamy virus-based vectors directing high level expression of foreign genes. Virology 210:167–178

Schnell T, Foley P, Wirth M, Munch J, Uberla K (2000) Development of a self-inactivating, minimal lentivirus vector based on simian immunodeficiency virus. Hum Gene Ther 11:439–447

Schweizer M, Falcone V, Gange J, Turek R, Neumann-Haefelin D (1997) Simian foamy virus isolated from an accidentally infected human individual. J Virol 71:4821–4824

Schweizer M, Turek R, Hahn H, Schliephake A, Netzer KO, Eder G, Reinhardt M, Rethwilm A, Neumann-Haefelin D (1995) Markers of foamy virus infections in monkeys, apes, and accidentally infected humans: appropriate testing fails to confirm suspected foamy virus prevalence in humans. AIDS Res Hum Retroviruses 11:161–170

Smith A (1999) Gene Therapy—where are we? Lancet 354:1–4

Sodroski J, Goh WC, Rosen C, Campbell K, Haseltine WA (1986) Role of the HTLV-III/LAV envelope in syncytium formation and cytopathicity. Nature 322:470–474

Soneoka Y, Cannon PM, Ramsdale EE, Griffiths JC, Romano G, Kingsman SM, Kingsman AJ (1995) A transient three-plasmid expression system for the production of high titer retroviral vectors. Nucleic Acids Res 23:628–633

Springett GM, Moen RC, Anderson S, Blaese RM, Anderson WF (1989) Infection efficiency of T lymphocytes with amphotropic retroviral vectors is cell cycle dependent. J Virol 63:3865–3869

Tobaly-Tapiero J, Bittoun P, Neves M, Guillemin MC, Lecellier CH, Puvion-Dutilleul F, Gicquel B, Zientara S, Giron ML, de The H, Saïb A (2000) Isolation and characterization of an equine foamy virus. J Virol 74:4064–4073

Tobaly-Tapiero J, Kupiec J-J, Santillana-Hayat M, Canivet M, Peries J, Emanoil-Ravier R (1991) Further characterization of the gapped DNA intermediates of human spumavirus: evidence for a dual initiation of plus-stranded DNA synthesis. J Gen Virol 72:605–608

Trobridge GD, Russel DW (1998) Helper-free foamy virus vectors. Hum Gene Ther 9:2517–2525

Trobridge G, Josephson N, Vassilopoulos G, Mac J, Russell DW (2002a) Improved foamy virus vectors with minimal viral sequences. Mol Ther 6:321–328

Trobridge G, Vassilopoulos G, Josephson N, Russell DW (2002b) Gene transfer with foamy virus vectors. Methods Enzymol 346:628–648

Varmus H (1988) Retroviruses. Science 240:1427–1435

Vassilopoulos G, Trobridge G, Josephson NC, Russell DW (2001) Gene transfer into murine hematopoietic stem cells with helper free foamy virus vectors. Blood 98:604–609

Venkatesh LK, Theodorakis PA, Chinnadurai G (1991) Distinct cis-acting regions in the U3 regulate trans-activation of the human spumaretrovirus long terminal repeat by the viral bel1 gene product. Nucl Acids Res 19:3661–3666

von-Laer D, Neumann-Haefelin D, Heeney J, Schweizer M (1996) Lymphocytes are the major reservoir for foamy viruses in peripheral blood. Virology 221:240–244

Weiss RA (1988) Foamy retroviruses. A virus in search of a disease. Nature 333:497–8

Winkler I, Bodem J, Haas L, Zemba M, Delius H, Flower R, Flügel RM, Löchelt M (1997) Characterization of the genome of feline foamy virus and its proteins shows distinct features different from those of primate spumaviruses. J Virol 71:6727–6741

Wolfe D, Goins WF, Yamada M, Moriuchi S, Krisky DM, Oligino TJ, Marconi PC, Fink DJ, Glorioso JC (1999) Engineering herpes simplex virus vectors for CNS applications. Exp Neurol 159:34–36

Wu M, Chari S, Yanchis T, Mergia A (1998) Cis-acting sequences required for simian foamy virus type 1 (SFV-1) vectors. J Virol 72:3451–3454

Wu M, Mergia A (1999) Packaging cell lines for simian foam virus type 1 (SFV-1) vectors. J Virol 73:4498–4501

Wu N, Ataai M (2000) Production of viral vectors for Gene Ther applications. Curr Opin Biotech 11:205–208

Yee JK, Friedmann T, Burns JC (1994) Generation of high-titer pseudotyped retroviral vectors with very broad host range. Methods Cell Biol 43:99–112

Yee JK, Miyanohara A, LaPorte P, Bouic K, Burns JC, Friedmann T (1994) A general method for the generation of high-titer, pantropic retroviral vectors: highly efficient infection of primary hepatocytes. Proc Natl Acad Sci USA 91:9564–9568

Yu H, Rabson AB, Kaul M, Ron Y, Dougherty JP (1996) Inducible human immunodeficiency virus type 1 packaging cell lines. J Virol 70:4530–4537

Yu SF, Baldwin DN, Gwynn SR, Yendapalli S, Linial ML (1996) Human foamy virus replication: a pathway distinct from that of retroviruses and hepadnaviruses. Science 271:1579–1582

Yu SF, Linial ML (1993) Analysis of the role of the bel and bet open reading frames of human foamy virus by using a new quantitative assay. J Virol 67:6618–6624

Yu SF, Sullivan MD, Linial ML (1999) Evidence that the human foamy virus genome is DNA. J Virol 73:1565–1572

Zennou V, Petit C, Guetard D, Nehrbass U, Montgnier L, Charneau P (2000) HIV-1 genome nuclear import is mediated by a central DNA flap. Cell 101:173–185

Zucali JR, Ciccarone T, Kelley V, Johnson CM, Mergia A (2002) Transduction of umbilical cord blood $CD34^+$ NOD/SCID repopulating cells by simian foamy virus type 1 (SFV-1) vector. Virology 302:229–235

Chapter 7

Replication of Primate Foamy Viruses in Natural and Experimental Hosts

V. Falcone, M. Schweizer, D. Neumann-Haefelin

Abstract Foamy viruses (FVs) are common apathogenic retroviruses readily spread by horizontal transmission in nonhuman primate and some other mammalian host populations. Primate FV infections have been known for half a century, i.e., 15 years before the definition of retroviruses and another 15 years before the detection of primate immune deficiency viruses. The emerging interest in human retroviruses

V. Falcone, D. Neumann-Haefelin (✉)
Department of Virology, Institute for Medical Microbiology and Hygiene,
University of Freiburg, Hermann-Herder-Straße 11, 79104 Freiburg, Germany
E-mail: nehfr@UKL.uni-freiburg.de

M. Schweizer
Department of Medical Biotechnology,
Paul Ehrlich Institute, Paul-Ehrlich-Straße 51, 63225 Langen, Germany

included primate FV, and although the role of human hosts for FV was greatly overestimated temporarily, enthusiastic researchers compiled invaluable data on molecular biology and classic as well as molecular epidemiology of these viruses. It has been shown that lytic FV infection in a wide range of cell cultures is in great contrast to the silent state of the infection in animals. Once transmitted by saliva via biting, FVs reside in all tissues as DNA copies, but their replication is untraceable except in oral submucosal cells, which are thought to supply the virus for transmission. FVs have not definitely been associated with any disease, regardless of viral phylogenetic differences. Various primate and nonprimate species have been used for studies on the natural carrier state and primary infection. Experimental infections have mostly proven to be inefficient in primates as well as lower laboratory animals. However, investigation of the immune response in FV-infected animals has only partly explained the control of FV replication in the animal host. Thus, the biological role of FV remains an enigma to be resolved in the future.

1 Introduction

Foamy virus (FV) infections of nonhuman primates have first been described in the 1950s, when a peculiar cytopathic effect (CPE) with formation of multinucleated and vacuolated giant cells was observed in primary kidney cell cultures from Old World monkeys of the *Macacaceae* family (Enders and Peebles 1954; Rustigian et al. 1955). The name foamy virus or spumavirus was derived from the foam-like appearance of syncytia in the infected monolayer cell cultures. The same CPE was also found on passage into permanent cell lines and diploid cell strains of primate and nonprimate origin. Research on primate FV greatly increased when they were classified as retroviruses (Scolnick et al. 1970; Parks et al. 1971) after detection of reverse transcriptase and even more when the existence of a human FV was suggested (Achong et al. 1971). However, after extreme speculations on FV prevalence and association with human diseases it has now been agreed that humans are not among the natural hosts of FV (Schweizer et al. 1995; Heneine et al. 1998; see also chapter by Heneine et al., this volume). Cytopathicity of FV infection in cell culture has no counterpart in histopathology, and in vivo, there has been no proof of FV-related pathogenesis up to now, neither in naturally and experimentally infected primates nor in accidentally in-

Table 1. Prototypes of simian foamy virus (SFV)

SFV serotype	Host species	Primate group	Reference
SFV-1	*Macaca cyclopsis* (Taiwan monkey)	Asian	Johnston 1961
Old World			
SFV-2	*Macaca mulatta* (Rhesus monkey)	Asian	Johnston 1961
Old World			
SFV-3	*Cercopithecus aethiops* (Recently changed to *Chlorocebus aethiops*) (African green monkey)	African	Stiles et al. 1964
Old World			
SFV-4	*Saimiri sciureus* (Squirrel monkey)	New World	Johnston 1971
SFV-5	*Galago cranicaudatus pangeniensis* (Bushbaby)	Prosimian	Johnston 1971
SFV-6	*Pan troglodytes* (Chimpanzee)	Ape	Hooks et al. 1972
SFV-7	*Pan troglodytes* (Chimpanzee)	Ape	Rogers et al. 1967
SFV-8	*Ateles spec.* (Spider monkey)	New World	Hooks et al. 1973
SFV-9	*Cebus spec.* (Capuchin monkey)	New World	Hooks and Gibbs 1975
SFV-10	*Papio cynocephalus* (Baboon)	African	Rhodes-Feuillette et al. 1979
Old World			
SFV-11	*Pongo pygmaeus* (Orangutan)	Ape	McClure et al. 1994

fected humans (Schweizer et al. 1997; see also chapter by Heneine et al., this volume).

2 Recovery of Foamy Virus From Nonhuman Primates

Markers of FV infection in nonhuman primates can, because of immunological cross-reactivity, readily be detected by serological assays, e.g., immunofluorescence tests or Western blots (Carski 1960; Fleming and Clarke 1970, Schweizer et al. 1995, 1997). Viral isolates have been obtained from prosimians, New World monkeys, and Asian and African Old world monkeys, as well as from African and Asian great apes (Table 1). It has been concluded that primate foamy viruses developed much earlier than primate lentiviruses, which naturally occurred only in African Old World monkeys and African apes before probable transmission to humans (Hirsch et al. 1995).

Isolation of FV from primates initially succeeded by cocultivation with permissive cell lines of tissue samples from various organs (Hooks and Gibbs 1975). Later it was found that the rate of viral recovery is also very high when simple throat swabs are used for inoculation of diploid cell cultures: In 50 of 60 attempts, isolates were obtained from 40 seropositive African green monkeys (AGMs). Furthermore, isolates could be obtained efficiently from infected primates by cocultivation of stimulated lymphocytes with diploid fibroblasts (Schweizer et al. 1997). It is important to use fibroblasts derived from a different species, because stimulation of lymphocytes by mitogens may result in induction of interferon-γ, which can hamper isolation in cell cultures from the same primate species (Falcone et al. 1999b). Traditionally, FV isolates were classified by serological tests (Hooks and Gibbs 1975). Eleven serotypes have been classified and called "simian" FV (SFV), although isolates from prosimians and great apes are included (see recent reviews by Lecellier and Saïb 2000; Meiering and Linial 2001). Present attempts at a new terminology (Meiering and Lineal 2001) may facilitate communication in the future but have not been finalized yet. The currently approved SFV serotypes are listed in Table 1. More recent isolates, e.g., one from gorilla (Bieniasz et al. 1995a), are not included in the list because definition of serotypes has been discontinued.

Nowadays, typing of FV isolates is usually achieved by molecular sequencing and sequence alignment. Various fragments of FV DNA (Mc-

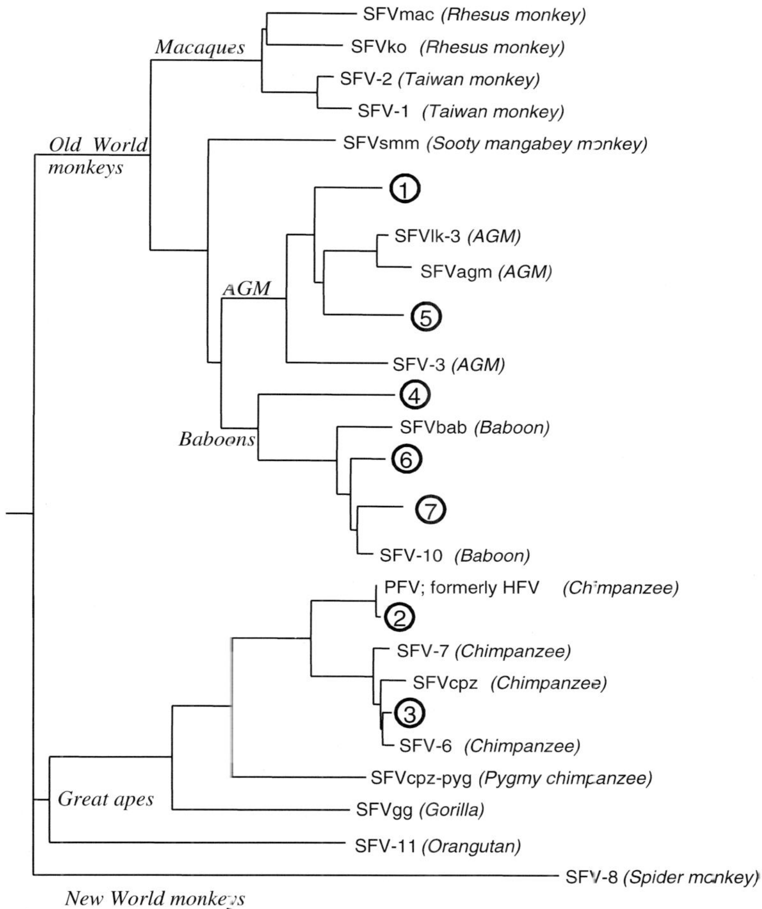

Fig. 1. Phylogenetic tree of primate FV prototype strains (SFV-1 to 11; see also Table 1), additional isolates from indicated nonhuman primate species, and viral nucleic acid sequences obtained from accidentally infected humans: (*1*) animal caretaker exposed to AGM (Schweizer et al. 1997); (*2*) laboratory worker exposed to PFV (Schweizer et al. 1995); (*3*) zoo worker exposed to chimpanzees (M. Schweizer, unpublished data); (*4–7*) primate center personnel exposed to various Old World monkeys (Heneine et al. 1998). The sequence comparison refers to *pol* (base pairs 5,994–6,418) of SFV-3 (Schweizer and Neumann-Haefelin 1995). Sequences were aligned with the CLUSTAL program, and the tree was compiled with the PHYLIP programs DNAdist and KITSCH. (Felsenstein 1989)

Clure et al. 1994; Broussard et al. 1997; Heneine et al. 1998) have been used for multiple sequence alignment. In our hands, a 425-bp region of the pol gene, flanked by highly conserved sequences, proved suitable for selection of PCR primers to compare foamy virus prototypes (Schweizer and Neumann-Haefelin 1995). As expected, the resulting phylogenetic tree reflects the evolutionary relationship of the host species investigated (Fig. 1). With this type of PCR, it was possible to classify new FV isolates or nucleic acids obtained from various primate species including humans (Schweizer et al. 1997; Heneine et al. 1998; Blewett et al. 2000). Thus molecular sequencing has proven to be the best tool for FV classification and may become the key of a new taxonomy.

3
Host Range of Primate Foamy Virus in Cell Culture

FVs are able to infect a wide range of cultured cells, comprising primary cells and cell strains or lines of epithelial, fibroblastic, myeloid and lymphoid origin (Hooks and Gibbs 1975). Moreover, the origin of permissive cells is not restricted to primate or even to mammalian species, because primate FVs have been propagated on cells derived from human, monkey, pig, dog, cat, a number of other mammals, and also quail, chicken, or iguana (Hooks and Gibbs 1975; McClure and Erlwein 1995). Indeed, no vertebrate cell line has been documented to be unequivocally uninfectable by FV. This broad host range in cell culture points to a cellular receptor which is ubiquitous on vertebrate cells. However, because a nonpermissive cell line is missing, the identification of a receptor molecule for FV is difficult.

The degree of permissiveness for FV replication varies in different cell types (Neumann-Haefelin et al. 1983). In many fibroblastic, epithelial, or neuronal cell lines, strong cytopathicity is induced which leads to rapid degeneration of infected cultures. The CPE is characterized by formation of large syncytia which contain cytoplasmic vacuoles, giving monolayers the "foamy" appearance. In some lymphoid cell suspension cultures, FV-infected cells look very similar to ballooned cells occurring in T-cell lines infected with syncytium-forming strains of human immunodeficiency virus (HIV). In other cases, in particular in several lymphoid cell lines, latent infection or low-level replication is observed which is characterized by absence or delayed occurrence of CPE and virus production (Yu et al. 1996; Mikovits et al. 1996; Mergia et al. 1996). Multiple studies investigating the molecular basis of different conditions of FV persistence in cell culture have been helpful in understanding the discrepancy between the strong cytopathogenicity in cell culture and the harmless course of infection in vivo (Schweizer et al. 1993; Bock et al. 1998; Callahan et al. 1999; Saïb et al. 1993, 1995a; Wybier-Franqui et al. 1995).

FV infection is cell cycle dependent: It has been shown that growth-arrested cells or cells progressing through a partial cycle but not completing mitosis cannot be infected (Bieniasz et al. 1995b). It was hypothesized that FVs, like C-type retroviruses, need dissolution of the nuclear membrane for integration and expression. However, studies on transduction efficiency for nondividing cells revealed some advantage of FV vectors over gammaretroviral vectors (Hirata et al. 1996, Mergia et al. 2001), thus leaving this interesting question open.

4 Primate Foamy Virus Persistence, Latency and Replication In Vivo

Even before classification as retroviridae, FVs were known to cause persistent infections in wild and laboratory animals. The virus can be isolated in monolayer cultures obtained from solid organs of infected animals or by cocultivation of peripheral blood mononuclear cells with permissive cells (Hooks and Gibbs 1975; Feldmann et al. 1975). Despite prolonged or life-long viral persistence, no confirmed report of associated disease has ever been documented. Primary B cells expanded from macaque blood using immortalization by Epstein-Barr virus (EBV)-like herpesvirus (Voss et al. 1992) were reported to develop the CPE typical of FV infection. FV-specific antigens were demonstrated in these cells, but it remained unclear whether the FV originated from the B-cells or other cell types present in the donor blood. A number of studies have tried to identify which subset of peripheral blood cells is persistently infected in FV-infected hosts. In a first study (von Laer et al. 1996) $CD8^+$ cells were reported as a major viral reservoir in AGM, chimpanzees and in two infected humans. All other human cell types analyzed ($CD4^+$, $CD14^+$ and $CD19^+$) were negative for FV DNA. On the contrary, in AGM samples, 4 of 7 $CD4^+$ samples, 10 of 13 $CD4^-$ $CD8^-$ (B cells) samples, 3 of 11 $CD14^+$ samples and 4 of 11 PMNL samples were found positive. A result contradictory to that obtained by von Laer et al. (1996) was reported in a recent study (Callahan et al. 1999). Here, only monocytes and B cells of another human infected with SFV-3 acquired from an

AGM were found to harbor viral DNA. Virus isolated from this individual showed a deletion in bel-2 ORF. Whether this deletion might explain the contradictory results has not been further investigated. A recent study has shown that naturally infected AGMs contain significant amounts of FV proviral DNA in all organs and that viral replication occurs at detectable levels only in a minimal number of cells of the oral mucosa (Falcone et al. 1999a). This finding is in agreement with unlimited oral FV shedding that makes biting the common mode of transmission of FV infection (Neumann-Haefelin et al. 1993) and presents for the first time direct evidence of viral transcripts in the host. No histopathological lesions could be observed in any of the organs analyzed; besides proviruses containing the complete transactivator gene, proviral genomes carrying a specific deletion in *tas* (△SFV-3 provirus) were detected in all monkeys. This is in analogy to the deletion described for PFV (Saïb et al. 1995a). It has been suggested that △PFV and Bet could be involved in the establishment of persistence. In fact, in the absence of Tas, Bet can be expressed from the internal promoter (IP) but no viral proteins can be expressed from the LTR promoter (P). However, the random distribution of both proviral forms reported in persistently infected AGM argues against a direct role for △PFV in the establishment of persistent infections. Moreover, a recent study reports the presence in an SFV-3-infected human of a virus lacking *bel2* ORF and consequently Bet (Callahan et al. 1999). This observation suggests that the virus is able to persist in vivo even if Bet is not produced and that absence of Bet does not augment viral titers. Some possible factors limiting viral replication in vivo have very recently been suggested (Linial 1999) such as different cellular factors required for Tas-driven transcription from P or IP, the role of Bet in preventing viral spread from permissive cells or a role of Bet in transcription downregulation. Another mechanism suppressing FV replication might be methylation of proviruses from which little transcription occurs (Schweizer et al. 1993).

5
Experimental Primary and Re-infection in Monkeys

AGMs are usually infected with SFV-3. Recent studies conducted by our group (E. Urbanyi, Veterinary thesis 1997) have shown that primary infection of AGMs with PFV, an evolutionarily distant FV of chimpanzee, is possible. In contrast to persistent natural infection, the course was that of an abortive infection. In detail, an intravenously infected AGM became seropositive 5 weeks after infection and FV DNA could be detected in lymphocytes 7 weeks p.i. over a period of 5 months. Afterwards, no DNA could be found anymore. Also, post-mortem analysis of various organs revealed no presence of proviral DNA although the animals remained seropositive. Repeated virus isolation attempts were unsuccessful, both from throat swabs or lymphocytes and from any other organs after sacrificing the animals. Low levels of neutralizing antibodies could be detected in the infected animals. Superinfection with PFV of SFV-3-infected AGMs was also possible but it was strictly dependent on the titer of the virus stock and the route of infection. In fact, only animals infected intramuscularly with a virus stock higher than 10^5 infectious units per milliliter revealed signs of PFV infection. PFV DNA could be detected in the lymphocytes and only in two organs (lymph nodes and skeletal muscle) of one of the re-infected animals. This finding reveals once more that the experimental infection is very inefficient as compared to natural transmission.

6
Foamy Virus Epidemiology and Impact on Natural Primate Hosts

Taking numerous reports together (Meiering and Linial 2001; Schweizer et al. 1988; 1995) it becomes evident that FV is not rare but prevalent at high rates in various great ape and monkey populations studied over more than three decades. Obviously, the adult population of clans or herds living in the wild normally harbor their proper type of FV. Babies and very young animals are usually found free of infection, and subsequent transmission appears to be associated with biting (Hooks and Gibbs 1975). Because excretion of virus and/or virus-infected cells via saliva does not depend on the stage of infection or the age of the virus carrier (Neumann-Haefelin et al. 1983), efficient early transmission may result from feeding or minor educational bites by the mothers, later by training and playing of adolescents, and finally by fighting of rivals. Sexual transmission of FV among monkeys cannot be excluded. However, it may be epidemiologically unimportant, because considerable rates of FV prevalence are observed before sexual maturity. Most of this information refers to Old World monkey and chimpanzee populations studied in breeding colonies of major primate centers and in animal houses of experimental settings (Blevett et al. 2000; Broussard et al. 1997; J. Heeney, unpublished data; D. Neumann-Haefelin, unpublished data). Much less is known on FV of New World monkeys (Johnston 1971; Hooks et al 1975), and their transmission to other primates has not been described. Recently, tamarins and marmosets living in a major primate center have been screened for FV prevalence without tracing any markers of infection (J. Heeney and E. Verschoor, unpublished data).

In monkeys, signs or symptoms of disease have never been associated with suspected or proven primary FV infection or conditions that might activate FV infections out of stable persistence or latency (reviewed by Linial 2000). A systematic survey by our group (J. Frey, Medical thesis 1985) conducted on viral excretion in AGMs included cyclosporin A-induced immunosuppression. It revealed only a slight enhancement of FV in saliva concomitant with shedding of EBV-like herpesvirus increased under this type of immunosuppression. No pathogenicity could be encountered beyond the known reversible side effects of cyclosporin A, mainly anorexia.

Apparently, transmission of FV is impaired by the presence of maternal antibodies in the fetus and the newborn. Thus it could not really be excluded that newborns and very young babies might be vulnerable to the virus if borne by seronegative mothers. Recent studies (J. Heeney and H. Nipheus, unpublished data) on this issue were initiated as a step towards safety assessment for exploitation of FV-based gene vectors. The few experiments evaluated so far did involve extreme difficulties in handling the newborn monkeys experimentally but did not suggest a definite risk for the babies, particularly due to the FV inoculum. The "type" of FV transmitted experimentally to seronegative monkeys does not seem to be crucial for the efficiency of infection. From a few transmissions, we rather had the impression that the route of inoculation might be of importance, with intramuscular injection being superior to the intravenous or intraperitoneal routes (V. Falcone, unpublished data). Cross-transmission of SFV-1, -2 and -3 seems to happen easily in primate centers (Hooks and Gibbs 1975), and even between monkeys and great apes there should be no major barriers. This has not been shown in particular studies but can be concluded from accidental human infections (see chapter by Heneine et al., this volume; Schweizer et al. 1995). Despite careful observation (Weiss 1988, 1996) no case of recorded inadvertent or experimental crossing of species borders by primate FV including zoonotic infections in humans (Heneine et al. 1998; Schweizer et al. 1995; Linial 2000) has given any hint to emerging pathogenicity.

7
Lower Animal Models for Primate Foamy Virus Infection

Several in vivo models have been established to increase the understanding of FV infection in the natural primate host. However, these studies have sometimes produced contradictory results regarding distribution of viral genomes in different organs, frequency of virus isolation and induction of pathological changes.

7.1
Rabbit

The group of Johnston (Johnston et al. 1974) reported that rabbits intradermally infected with SFV-1 show inflammation and necrosis at the site of injection; at this site, as well as from kidney explant cultures, very low titers of virus could be isolated up to 1 year postinfection. In another study, intraperitoneal and intranasal infection of New Zealand white rabbits with SFV-1 has been shown to establish FV persistence even in the presence of significant levels of neutralizing antibody (Swack and Hsiung 1975). Virus could be recovered from all organs analyzed (264 days p.i. from spleen, liver, lung, salivary gland, kidney) and to a lesser extent from the brain. Virus isolation from blood cells was successful only during the early phase of infection. Distribution of FV in naturally infected rhesus monkeys and baboons was shown to be similar to that observed in experimentally infected rabbits. Inoculation of rabbits with SFV-7 has also been reported to produce a persistent infection (Hooks and Detrick-Hooks 1979). However, in this case virus could be isolated from peripheral blood mononuclear cells only for 6 months after inoculation. In this study a unique hint to pathology associated with foamy virus was first described. In fact, the lymphocytic response of infected rabbits to phytohemagglutinin stimulation in vitro was depressed only transiently during the first 2 weeks after infection in the absence of interferon production. More recently, a transient suppression of the cell-mediated immune response was also reported in rabbits intravenously infected with PFV (Santillana-Hayat et al 1993). Nevertheless, the absence of progression to an immunodeficient state in both studies suggests that this phenomenon has no clinical relevance but could play a role in determining viral persistence.

PFV has also been shown to establish a persistent infection in subcutaneously infected domestic rabbits (Saïb et al. 1997). In this study as well, no pathology was observed but the animals developed a strong immunological response against major viral proteins. PFV could be rescued by cocultivation of PBLs with permissive cells 2 weeks after infection. By PCR, all organs tested were positive for PFV DNA. Moreover, the defective form of PFV containing a deletion of 301 bp in *tas*, generated by alternative splicing of the pregenomic RNA (Saïb et al. 1995a), was shown to be the predominant form present in PBL and organs of persistently infected rabbits, suggesting that in vivo viral persistence could be related to homologous interference between △PFV and its wild form, as previously shown in vitro (Saïb et al. 1995b).

7.2 Mouse

The group of Gajdusek tried to establish a model for FV infection in the mouse (Brown et al. 1982). In Swiss-Webster mice, infection with SFV-6 (originally isolated from the kidney of a kuru-inoculated chimpanzee) remained asymptomatic, but virus could be detected in explant cultures made from kidney, spleen and brain for up to 10 months after intraperitoneal infection; in this study the presence of complement fixing but not neutralizing antibodies is reported. The use of modern molecular biological techniques has provided new tools to study the interactions of FV with its host. Schmidt and colleagues (Schmidt et al. 1997) have established a mouse model using molecularly cloned PFV. CBA and C57BL mice intraperitoneally infected with PFV developed a persistent infection in the absence of clinical symptoms. Whereas PFV DNA could be detected by PCR in virtually all organs of CBA mice, only a few organs of C57BL tested positive. The different course of infection in the two mouse strains reflects differences in the antiviral immune response (see below). In this study, the frequency of virus re-isolation was very low (2 of 52 infected CBA mice). The use of defined molecularly cloned virus, the methods used for virus isolation and the genetic background of the mice might explain the difference in results between these studies and those of Brown et al. (1982). Interestingly, PFV derived from molecular clones induced a lethal disease in athymic mice after an incubation period of 9 to 30 weeks (Imrich et al. 2003):

7.3
PFV-Transgenic Mouse

PFV-transgenic mice represent the only animal model showing that FV, besides causing cytopathic effects in cultured cells, can induce pathology in vivo. Two transgenic strains of mice were generated (Bothe et al. 1991). The first expressed all genes of PFV from the LTR but was replication-defective because of a frameshift in pol. The second strain expressed only the accessory genes *tas* and *bet*. In both strains the animals were shown to develop a severe neurological syndrome at 6–8 weeks of age leading to death 4–6 weeks from onset. Disease progression was faster in the strain expressing all FV genes. In all animals the pathological findings were restricted to the CNS and the striated muscle. Most frequently the CA3 layer of the hippocampus and the telencephalic cortex were affected (Bothe et al. 1991). Atypical multinucleated giant cells containing 5–10 nuclei were often found in the brain of transgenic mice expressing all structural PFV genes. Immunostaining of these cells revealed that they contain *Gag*, *Env* and *Tas*. The gene products responsible for pathology were not identified, but it was likely to be *tas* and *bet*, the only two intact ORFs in both constructs. The results of the majority of theses studies are reviewed in detail elsewhere (Aguzzi 1993; Aguzzi et al. 1996). Ensuing speculations on a direct correlation between FV and encephalopathy in an orangutan (McClure et al. 1994) and neurological disorders in humans (Westarp et al. 1993, Cameron et al. 1978) have been challenging, but a causal link between FV infection and such diseases has never been proven (Rösener et al. 1996; Schweizer et al. 1995).

8
Immune Response to Primate Foamy Virus

The immune response during FV infections is not well characterized. Natural and experimental infections in primates induce antibodies mainly to Gag and Bet (Hahn et al. 1994). Low titers of neutralizing antibodies can be detected (Hooks and Gibbs 1975; Swack and Hsiung 1975; Neumann-Haefelin et al. 1983) but their role in controlling viral spread in vivo is controversial. In the 1970s Hooks and colleagues showed that FVs are able to spread from cell to contiguous cell despite presence of neutralizing antibodies (Hooks et al. 1976). The presence of complement-fixing but not neutralizing antibodies has been reported in SFV-6-

infected mice (Brown et al. 1982). A strong anti-Gag and anti-Bet response has been described in C57BL mice experimentally infected with PFV as proven by Western blot in naturally infected primates. In CBA mice, a weaker anti-Gag response and no detectable anti-Bet response was reported. Only few CBA mice harbored detectable levels of FV DNA in the early phase of infection, but over the time, the number of positive animals and positive organs increased. The opposite situation was observed in C57BL/6 mice: The number of FV DNA-positive animals and organs gradually decreased during the course of infection. The authors concluded that the Bet protein may play a crucial role in spreading the virus or that an immune reaction against Bet is essential in controlling virus spread (Schmidt et al. 1997).

The failure of antiviral antibody and complement to lyse infected cells and the poor induction of and relative resistance to interferons (IFNs) have been regarded as possible mechanisms contributing to persistence (Hooks et al. 1976). Later studies by Rhodes-Feuillette confirmed that SFV-1 and -2 do not induce IFN in mouse and primate cell lines and showed that the characteristic CPE induced by FV can be blocked by IFN (Rhodes-Feuillette et al. 1987). The same group also showed the inhibitory effect of human recombinant IFN on the replication of SFV1 and PFV in human AV3 cell cultures. The strong inhibition of the FV CPE is accompanied by a 70%–95% decrease in viral particles and in a decrease of RT activity in supernatants of IFN-treated cultures (Rhodes-Feuillette et al. 1990; Sabile et al. 1996). Recently, Regad et al. (2001) have tested the capacity of three IFN mediators, the promyelocytic leukemia gene (PML), MxA and Mx1, to interfere with PFV replication. Overexpression of PML, but not of MxA and Mx1, confers a high resistance to PFV replication by inhibiting viral antigen expression at the transcription level. PML interacts directly with Tas and interferes with its ability to bind PFV LTR and IP, leading to the repression of Tas-induced transactivation and inhibition of PFV in PML-expressing and in IFN-treated cells. In a previous study (Falcone et al. 1999b) our group had shown that activated peripheral blood lymphocytes produce a soluble factor inhibiting the SFV-3-induced CPE in a species-specific manner. IFN-γ is the major component responsible for the FV-suppressive activity; however, because the suppressive activity is not entirely blocked by anti-IFN-γantibodies, other soluble factors might act in concert with IFN-γ. Whether IFNs also play a protective role against lytic replication of FV in vivo still remains to be clarified. However, it might be that, even if

FVs are poor inducers of IFN in cell culture, indirect mechanisms can induce sufficient amounts of these cytokines.

Several viruses including adenovirus (Kralli et al. 1992), human T-lymphotropic virus (Sawada et al. 1990) and HIV (Howcroft et al. 1993) have been shown to modulate host MHC expression levels. The expression of MHC-I genes in U373-MG cells is increased on infection with PFV; this effect is not mediated by the IFN pathway (Colas et al. 1995). The role of the modulation of MHC gene expression in PFV-infected cells has to be elucidated, although the possible involvement in the induction of autoimmune diseases has never been confirmed. The T-cell response to FV has not yet been studied. However, the role of $CD4^+$ and $CD8^+$ T cells may be highly important in controlling viral replication. Thus it might be significant, especially in the context of the other primate retrovirus infections, to understand the early events of FV infection leading from initial replication at the site of entry to latency in multiple organs of the host.

9
Conclusions and Perspectives

Most nonhuman primates acquire asymptomatic FV infection rather early in life, mount an immune response, and remain lifelong viral carriers and transmitters, obviously without any advantage or disadvantage related to this infection. The rare accident of transmission to humans seems even less eventful in being a dead end for the virus. Thus FV would appear medically unimportant but for the unanswered questions about the past evolution and certain speculations on the future.

One is tempted to ask why this remote phenomenon traceable only by laboratory means as cell culture, polymerase chain reaction and artificial immune reactions is endemic in the majority of primate species. Why does it exist as an "orphan" virus "in search of a disease" and why does it spare humans as the only–or one among a few–primate species?

One may wonder, where FV stands in evolution compared to endogenous retroviruses at one end and HIV at the other end of the scale. It might be asked how many nucleotide exchanges are required to make FV pathogenic and able to spread beyond their natural host species, e.g., in human populations; stated the other way round, what is essential to keep FV so perfectly host-adapted? Their potential of getting out of control has been suggested so far only in one rather artificial (transgenic) animal model, and it is hard to see what this model could stand for.

Moreover, it has been suggested that FVs might serve as biological tools after being transformed from innocent passenger to spy or Trojan horse. However, potential benefit and risk may be close together in such ambitious visions. In any event, efforts on this line seem to be limited primarily because of the inert nature of FV. A more advanced understanding of FV interaction with the natural host will be at least a prerequisite, and might become the key, for potential exploitation of this particular retroviral genus.

Acknowledgements. We thank Jonathan Heeney for access to unpublished data and approval of using them for this review and Gerhard Hunsmann as well as Christiane Stahl-Hennig for helpful advice and training of personnel for experiments with non-human primates. Andreas Schmeel (Chiron Behring) and colleagues in the university of Freiburg, Norbert Boehm (Pathology), Benedikt Volk (Neuropathology), and Dr. Wolfbernhard Spatz (Neurophysiology) generously contributed to our own studies in African green monkeys that were supported by the Deutsche Forschungsgemeinschaft (Ne 213/4-4 and Ne 213/5-1) and the European Commission (BMH4-CT97-2010). The secretarial assistance of Brunhilde Strohecker is gratefully appreciated.

References

Achong BG, Mansell PWA, Epstein MA, Clifford P (1971) An unusual virus in cultures from a human nasopharyngeal carcinoma. J Natl Cancer Inst 46:299–302

Aguzzi A (1993) The foamy virus family: molecular biology, epidemiology and neuropathology. Biochim Biophys Acta 1155:124

Aguzzi A, Marino S, Tschopp R, Rethwilm A (1996) Regulation of expression and pathogenic potential of human foamy virus in vitro and in transgenic mice. Curr Top Microbiol Immunol 206:243273

Bieniasz PD, Rethwilm A, Pitman R, Daniel MD, Chrystie I, McClure MO (1995a) A comparative study of higher primate foamy viruses, including a new virus from a gorilla. Virology 207:217–228

Bieniasz PD, Weiss RA, McClure MO (1995b) Cell cycle dependence of foamy retrovirus infection. J Virol 69: 7295–7299

Blewett EL, Black DH, Lerche NW, White G, Eberle R (2000) SFV infection in a baboon breeding colony. Virology 278:183–193

Bock M, Heinkelein M, Lindemann D, Rethwilm A (1998) Cells expressing the human foamyvirus (HFV) accessory bet protein are resistant to productive HFV superinfection. Virology 250:194–204

Bothe K, Aguzzi A, Lassmann H, Rethwilm A, Horak I (1991) Progressive encephalopathy and myopathy in transgenic mice expressing human foamy virus genes. Science 253:555–557

Broussard SR, Comuzzie AG, Leighton KL, Leland MM, Whitehead EM, Allan JS (1997) Characterization of new simian foamy viruses from African nonhuman primates. Virology 237: 349–59

Brown P, Morreau-Dubois C, Gajdusek DC (1982) Persistent asymptomatic infection of the laboratory mouse by simian foamy virus type 6: a new model of retroviral latency. Arch Virol 71: 229–234

Callahan ME, Switzer WM, Matthews AL, Roberts BD, Heneine W, Folks TM, Sandstrom PA (1999) Persistent zoonotic infection of a human with a simian foamy virus in the absence of an intact orf-2 accessory gene. J Virol 73:9619–24

Cameron, KR, Birchall SM, Moses, MA (1978) Isolation of foamy virus from patient with dialysis encephalopathy. Lancet II:796

Carski T (1960) A fluorescent antibody study of the simian foamy agent. J Immunol 84: 426–433

Colas S, Bourge JF, Wybier J, Chelbi-Alix MK, Paul P, Emanoil-Ravier R (1995) Human foamy virus infection activates class I major histocompatibility complex antigen expression. J Gen Virol 76:661–667

Enders JF, Peebles TC (1954) Propagation in tissue cultures of cytopathogenic agents from patients with measles. Proc Soc Exp Biol Med 86:277–286

Falcone V, Leupold J, Clotten J, Urbanyi E, Herchenroeder O, Spatz W, Volk B, Boehm N, Toniolo A, Neumann-Haefelin D, Schweizer M (1999a) Sites of simian foamy virus persistence in naturally infected African green monkeys: latent provirus is ubiquitous, whereas viral replication is restricted to the oral mucosa. Virology 257:7–14

Falcone V, Schweizer M, Toniolo A, Neumann-Haefelin D, Meyerhans A (1999b) Gamma interferon is a major suppressive factor produced by activated human peripheral blood lymphocytes that is able to inhibit foamy virus-induced cytopathic effects. J Virol 73:1724–1728

Feldman MD, Dunnick NR, Barry DW, Paerkman PD (1975) Isolation of foamy virus from rhesus, African green and cynomolgus monkey leukocytes. J Med Primatol 4:287–295

Felsenstein J (1989) PHYLIP (phylogeny inference package) version 3.2. Cladistics 5:164–166

Fleming WA, Clarke JK (1970) Fluorescence assay of foamy virus. J Gen Virol 6:277–284

Hahn H, Baunach G, Bräutigam S, Mergia A, Neumann-Haefelin D, Daniel MD, McClure M, Rethwilm A (1994) Reactivity of primate sera to foamy virus Gag and Bet proteins. J Gen Virol 75:2635–2644

Heneine W, Switzer WM, Sandstrom P, Brown J, Shanmugam V, Schable CA, Khan AS, Lerche NW, Schweizer M, Neumann-Haefelin D, Chapman LE, Folks, TM (1998) Identification of a human population infected with simian foamy virus. Nat Med 4:403–407

Hirata RK, Miller AD, Andrews RG, Russell DW (1996) Transduction of hematopoietic cells by foamy virus vectors. Blood 88:3654–61

Hirsch VM, Dapolito G, Goeken R, Campbell BJ (1995) Phylogeny and natural history of the primate lentiviruses, SIV and HIV. Curr Opin Genet Dev 5:798–806

Hooks JJ, Burns W, Hayashi K, Geis S, Notkins AL (1976) Viral spread in the presence of neutralizing antibody: mechanisms of persistence in foamy virus infection. Infect Immun 14:1172–1178

Hooks JJ, Gibbs CJ, Chou R, Howk R, Lewis M, Gajdusek DC (1973) Isolation of a new simian foamy virus from a spider monkey brain culture. Infect Immun 8:804–813

Hooks JJ, Gibbs CJ, Cutchins EC, Rogers NG, Lampert P, Gajdusek DC (1972) Characterization and distribution of two new foamy viruses isolated from chimpanzees. Arch Ges Virusforsch 38:38–55

Hooks JJ, Gibbs CJ (1975) The foamy viruses. Bacteriol Rev 39:169–185

Hooks JJ, Detrick-Hooks B (1979) Simian foamy virus-induced immunosuppression in rabbits. J Gen Virol 44: 383–390

Howcroft TK, Strebel K, Martin MA, Singer DS (1993) Repression of MHC-1 gene promoter activity by two-exon tat of HIV. Science 260:1320–1322

Imrich H, Müller JG, Lindemann D, Heinkelein M, Rethwilm A (2003) Pathogenicity of Foamy Virus in arhymic mice. (Submitted)

Johnston PB (1961) A second immunological type of simian foamy virus: monkey throat infection and unmasking by both types. J Infect Dis 109:1–9

Johnston PB (1971) Taxonomic features of seven serotypes of simian and ape foamy viruses. Infect Immun 3:793–799

Johnston PB (1974) Studies on simian foamy viruses and syncytium-forming viruses of lower animals. Lab Anim Sci 24:159166

Kralli A, Ge R, Graeven U, Ricciardi RP, Weinmann R (1992) Negative regulation of the major histocompatibility complex class I enhancer in adenovirus type 12-transformed cells via a retinoic acid response element. J Virol 66:6979–6988

Lecellier CH, Saïb A (2000) Foamy viruses: Between retroviruses and pararetroviruses. Virology 271:1–8

Linial M (2000) Why aren't foamy viruses pathogenic? Trends Microbiol 289:284–289

Linial ML (1999) Foamy viruses are unconventional retroviruses. J Virol 73:17471755

McClure MO, Bieniasz PD, Schulz TF, Chrystie IL, Simpson G, Aguzzi A, Hoad JG, Cunningham A, Kirkwood J, Weiss RA (1994) Isolation of a new foamy retrovirus from orangutans. J Virol 68:7124–7130

McClure MO, Erlwein O (1995) Foamy viruses—pathogenic or therapeutic potential. Rev Med Virol 5: 229–237

Meiering CD, Linial ML (2001) Historical perspective of foamy virus epidemiology and infection. Clin Microbiol Rev 14:165–176

Mergia A, Chari S, Kolson DL, Goodenow MM, Ciccarone T (2001) The efficiency of simian foamy virus vector type-1 (SFV-1) in nondividing cells and human PBLs. Virology 280:243–252

Mergia A, Leung NJ, Blackwell J (1996) Cell tropism of the simian foamy virus type 1 (SFV-1). J Med Primatol 25:2–7

Mikovits JA, Hoffman PM, Rethwilm A, Ruscetti FW (1996) In vitro infection of primary and retrovirus-infected human leukocytes by human foamy virus. J Virol 70: 2774–2780

Neumann-Haefelin D, Fleps U, Renne R, Schweizer M (1993) Foamy viruses. Intervirology 35:196–207

Neumann-Haefelin D, Rethwilm A, Bauer G, Gudat F, zur Hausen H (1983) Characterization of a foamy virus isolated from *Cercopithecus aethiops* lymphoblastoid cells. Med Microbiol Immunol 172:75–86

Parks WP, Todaro GJ, Scolnick EM, Aaronson SA (1971) RNA dependent DNA polymerase in primate syncytium-forming (foamy) viruses. Nature 229:258–260

Regad T, Saïb A, Lallemand-Breitenbach V, Pandolfi PP, de Thé H, Chelbi-Alix MK (2001) PML mediates the interferon-induced antiviral state against a complex retrovirus via its association with the viral transactivator. EMBO J 20:3495–3505

Rhodes-Feuillette A, Lasneret J, Paulien S, Ogunkolade W, Périès J, Canivet M (1990) Effects of human recombinant alpha and gamma and that of highly purified natural beta interferons on simian spumavirinae prototype (simian foamy virus 1) multiplication in human cells. Res Virol 141:31–43

Rhodes-Feuillette A, Saal F, Lasnaret J, Dubouch P, Périès J (1979) Isolation and characterization of a new simian foamy virus serotype from lymphocytes of a *Papio cynocephalus* baboon. J Med Primatol 8:308–320

Rhodes-Feuillette A, Saal F, Lasneret J, Santillana-Hayat M, Périès J (1987) Studies on in vitro interferon induction capacity and interferon sensitivity of simian foamy viruses. Arch Virol 97:77–84

Rogers N, Basnight M, Gibbs GJ, Gajdusek DC (1967) Latent viruses in chimpanzees with experimental kuru. Nature 216:446–449

Rösener M, Hahn H, Kranz M, Heeney J, Rethwilm A (1996) Absence of serological evidence for foamy virus infection in patients with amyotrophic lateral sclerosis. J Med Virol 48:222–226

Rustigian R, Johnston P, Reihart, H (1955) Infection of monkey kidney tissue cultures with virus-like agents. Proc Soc Exp Biol Med 88:8–16

Sabile A, Rhodes-Feuillette A, Jaoui FZ, Tobaly-Tapiero J, Giron ML, Lasneret J, Périès J, Canivet M (1996) In vitro studies on interferon-inducing capacity and sensitivity to IFN of human foamy virus. Res Virol 147:29–37

Saïb A, Koken HM, van der Spek P, Périès J, de Thé H (1995b) Involvement of a spliced and defective human foamy virus in the establishment of chronic infection. J Virol 69:5261–5268

Saïb A, Neves M, Giron ML, Guillemin MC, Valla J, Périès J, Canivet M (1997) Long-term persistent infection of domestic rabbits by the human foamy virus. Virology 228:263–268

Saïb A, Périès J, de Thé H (1995a) Recent insights into the biology of the human foamy virus. Trends Microbiol:173–178

Saïb A, Périès J, de Thé H (1993) A defective human foamy provirus generated by pregenome splicing. EMBO J 12:4439–4444

Santillana-Hayat M, Rozain F, Bittoun P, Chopin-Robert C, Lasneret J, Périès J. Canivet M (1993) Transient immunosuppressive effect induced in rabbits and mice by the human spumaretrovirus prototype HFV (human foamy virus). Res Virol 144:389–396

Sawada M, Suzumura A, Yoshida M, Marunouchi T (1990) Human T-cell leukemia virus type 1 transactivator induces class I major histocompatibility complex antigen expression in glial cells. J Virol 64:4002–4006

Schmidt M, Niewiesk S, Heeney J, Aguzzi A, Rethwilm A (1997) Mouse model to study the replication of primate foamy viruses. J Gen Virol 78: 1929–1933

Schweizer M, Corsten B, Neumann-Haefelin D (1988) Heterogeneity of primate foamy virus genomes. Arch Virol 99:125–134

Schweizer M, Falcone V, Gaenge J, Turek R, Neumann-Haefelin D (1997) Simian foamy virus isolated from an accidentally infected human individual. J Virol 71:4821–4824

Schweizer M, Fleps U, Jaeckle A, Renne R, Turek R, Neumann-Haefelin D (1993) Simian foamy virus type 3 (SFV-3) in latently infected Vero cells: Reactivation by demethylation of proviral DNA. Virology 192:663–666

Schweizer M, Neumann-Haefelin D (1995) Phylogenetic analysis of primate foamy viruses by comparison of pol sequences. Virology 207:577–582

Schweizer M, Schleer H, Pietrek M, Liegibel J, Falcone V, Neumann-Haefelin D (1999). Genetic stability of foamy viruses: long-term study in an African green monkey population. J Virol 73:9256–9265

Schweizer M, Turek R, Hahn H, Schliephake A, Netzer KO, Eder G, Reinhardt M, Rethwilm A, Neumann-Haefelin D (1995) Markers of foamy virus (FV) infections in monkeys, apes, and accidentally infected humans: appropriate testing fails to confirm suspected FV prevalence in man. AIDS Res Hum Retrovir 11:161–170

Scolnick E, Rands E, Aaronson SA, Todaro GJ (1970) RNA-dependent DNA polymerase activity in five RNA viruses: Divalent cation requirements. Proc Natl Acad SCI USA 67:1789–1796

Stiles GE, Bittle JL, Cabasso UJ (1964) Comparison of simian foamy virus strains including a new serological type. Nature 201:1350–1351

Swack NS, Hsiung GD (1975) Pathogenesis of simian foamy virus infection in natural and experimental hosts. Infect Immun 12:470–474

von Laer D, Neumann-Haefelin D, Heeney JL, Schweizer M (1996) Lymphocytes are the major reservoir for foamy viruses in peripheral blood. Virology 221:240–244

Voss G, Nick S, Stahl-Hennig C, Ritter K, Hunsmann G (1992) Generation of macaque B-lymphoblastoid cell lines with simian Epstein-Barr-like viruses: transformation procedure, characterization of the cell lines and occurrence of simian foamy virus. J Virol Methods 39:185–195

Weiss RA (1988) Foamy retroviruses. A virus in search of a disease. Nature 333:497–498

Weiss RA (1996) Foamy viruses bubble on. Nature 380: 201

Westarp ME, Fuchs D, Bartmann P, Hoff-Jorgensen R, Clausen J, Wachter H, Kornhuber HH (1993) Amyotrophic lateral sclerosis: an enigmatic disease with B-cellular and anti-retroviral immune responses. Eur J Med 2:327–332

Wybier-Franqui J, Tobaly-Tapiero J, Coronel A, Giron ML, Chopin-Robert C, Périès J, Emanoil-Ravier R (1995) Human foamy virus DNA forms and expression in persistently infected Dami megakaryocytic cells. AIDS Res Hum Retrovir 11:829–36

Yu SF, Stone J, Linial ML (1996) Productive persistent infection of hematopoietic cells by human foamy virus. J Virol 70:1250–1254

Chapter 3

Human Infection with Foamy Viruses

W. Heneine, M. Schweizer, P. Sandstrom, T. Folks

Abstract Virtually all nonhuman primate species investigated thus far including prosimians, New World and Old World monkeys and apes all harbor distinct and species-specific clades of simian foamy virus (SFV). However, evidence supporting the existence of a human-specific foamy virus (FV) is not yet available. Early reports describing widespread infection of healthy and sick humans with FV could not be confirmed. In contrast, all FV infections documented in humans are of zoonotic origin

W. Heneine (✉), T. Folks
HIV and Retrovirology Branch, Centers for Disease Control and Prevention, 1600 Clifton Road, Mail Stop G19, Atlanta, GA 30333, USA
E-mail: Wmh2@cdc.gov

M. Schweizer
Paul-Ehrlich-Institut, 63225 Langen, Germany

P. Sandstrom
Bureau of HIV/AIDS, STD and TB, Health Canada, Ottawa, Canada

and are identified in persons occupationally exposed to nonhuman primates. The introduction of SFV into humans raises several public health questions regarding disease outcomes and potential for human-to-human transmissibility. The available data from a very limited number of SFV-infected humans suggest that these infections are nonpathogenic and are not easily transmissible. Additional studies are needed to better define the prevalence and natural history of SFV in humans.

1 Introduction

A foamy virus (FV) was isolated from an African patient with nasopharyngeal carcinoma in 1971 before the discovery of either the human T-cell lymphotropic virus type 1 (HTLV-1) or the human immunodeficiency virus type 1 (HIV-1) (Achong et al. 1971). This viral isolate, previously designated as human syncytial virus, human syncytium-forming virus, human spumaretrovirus, or human foamy virus, has recently been renamed the prototype foamy virus (PFV). Although this report may represent the first evidence of a retrovirus detected in a human, several studies have questioned whether PFV represents an endemic infection among humans (Nemo et al. 1978; Schweizer et al. 1995). More recent work has shown that PFV is phylogenetically related to chimpanzee-type simian foamy viruses (SFVs), suggesting that it is of chimpanzee origin (Herchenroder 1994). To date, it remains uncertain whether PFV represents a genuine FV isolate from a cross species-infection or a laboratory contaminant accidentally introduced during the reportedly prolonged isolation process. Regardless of its origins, PFV has been the most intensely studied member of the FV genus over the past 25 years, responsible for most of our current understanding of the virology and genetics of FVs. An intensive search for FV infection in different human populations followed the first report of PFV (Weiss 1988) and has led to many reports of FV infection in both healthy and sick humans. The merit of these reports is discussed below.

2 Laboratory Diagnosis of FV Infection

Early studies looking at the prevalence of FV in humans were frequently fraught with the use of diagnostic methods that were inadequately validated or interpreted. It is therefore important to highlight the various diagnostic methods for which a reasonable validation has been reported. For antibody testing, two such serologic assays are available. These include immunofluorescence assays with PFV- or SFV-infected cells and a Western blot (WB) assay using whole antigens from PFV or other SFV strains. Both assays have been validated on sera from nonhuman primates (NHP) whose infection status has been documented by virus isolation and/or polymerase chain reaction (PCR) testing (Neumann-Haefelin et al. 1993; Hahn et al. 1994; Schweizer et al. 1995). Criteria for seropositivity were defined as reactivity to the gag precursor proteins of 71 and 68 kDa and nuclear staining of PFV-infected cells in the immunofluorescent assay. Cytoplasmic fluorescence appearing in giant cells without any evidence of nuclear staining has been found to be nonspecific and is usually not associated with WB positivity (Schweizer et al. 1995). ELISA assays with whole virus lysates have also been used successfully for antibody screening (McClure and Erlwein 1995; Ali et al. 1996). Because limited serologic cross-reactivity may exist between divergent SFV variants, consideration needs to be given to the strains of SFV used as source antigens in the serologic assays. The use of antigen from one strain of SFV may result in diminished sensitivity for detecting antibodies to divergent SFV variants from distantly related NHP species origin. Therefore, for better detection of antibodies to a wide spectrum of SFV variants it is advisable to test with more than one SFV antigen representing distantly related viruses, as previously done with the use of both chimpanzee and African green monkey type SFV antigens (Schweizer et al. 1995).

PCR analysis of peripheral blood lymphocytes can detect FV sequences both in infected humans and NHP and is considered an important diagnostic method (Schweizer et al. 1995; von Laer et al. 1996; Heneine et al. 1998). Both PFV-specific primer pairs and generic primer pairs based on conserved sequences among SFV from Old World monkeys have been successfully used to amplify FV of different NHP species (Schweizer et al. 1995; Heneine et al. 1998). However, the risk of false-positive results in PCR testing due to contamination is well recognized, and therefore, PCR positivity alone may not be sufficient evidence of infection.

Virus isolation also provides a useful tool for the diagnosis of FV infection in humans. FV are highly cytopathic in many types of cells, leading to rapid syncytium formation and vacuolization of cells. Baby hamster kidney (BHK-21) cells are particularly sensitive to FV-induced cytopathic effects. SFV has been successfully isolated from PBLs of SFV-infected humans by cocultivation with many cell types including canine thymocytes, Mus dunni, Raji, and A204 cells (Heneine et al. 1998). FV replication in these cells is associated with cytopathic effects, increase in reverse transcriptase activity, and PCR positivity for proviral sequences (Heneine et al. 1998). Because the risk of contamination of virus cultures is well known, virus isolation of FV from humans should not constitute the only evidence of infection but should be used in addition to data from serologic and PCR testing.

3 Seroprevalence in Human Populations from Different Regions

Although some early studies have found positive serologic results and prevalence rates of PFV that exceed 10% in different populations including cancer patients from Africa, HIV-1-positive persons in Tanzania, and healthy populations in the Solomon Islands (Loh et al. 1980; Achong and Epstein 1978; Muller et al. 1980; Mahnke et al. 1992; Achong and Epstein 1983), other studies failed to detect any seropositivity and suggested that previously found serologic reactivity was nonspecific (Nemo et al. 1978; Brown et al. 1978). Many technical reasons, such as the interpretation of cytoplasmic staining as positive immunofluorescent results, limited assay validation, and the lack of confirmatory testing, may explain the false-positive results reported in some studies. Large serosurveys done with validated assays in 1995 in more than 2,000 sera from populations previously suspected to have high FV prevalence showed no evidence of seropositivity (Schweizer et al. 1995). A British study of more than 6,000 sera from a global collection of sera including samples from the Solomon Islands also failed to detect specific antibody reactivity to PFV (McClure and Erlwein 1995; Ali et al. 1996). These findings led to the general consensus that infection with PFV-related viruses is not as prevalent among humans as previously thought.

4 Association with Human Disease

Although many reports have in the past associated FV infection with different human diseases, there is to date no convincing evidence that FVs are pathogenic in humans. Many of the disease associations could not be independently confirmed because clear evidence of FV infection in the patients studied could not be reproduced. The following is a description of the human diseases in which an FV etiology has been examined. Nasopharyngeal carcinoma (NPC) was studied because of the isolation of PFV from a Kenyan patient with NPC (Achong et al. 1971) and the subsequently reported high seroprevalence in African NPC patients (Achong and Epstein 1978; Muller et al. 1980). However, sera from NPC patients from different regions (Tunisia, Singapore, Great Britain) tested negative in these studies, indicating that an association of PFV with NPC is not supported. Studies of patients with de Quervain thyroiditis in Slovenia reported virus isolation from different organs (Stancec and Gressnerová 1974; Stancec et al. 1975; Stancekova et al. 1976; Werner and Gelderblom 1979). FV infection was suggested by immunofluorescence and neutralization tests. However, the association with PFV infection was refuted in more recent studies (Debons-Guillemin et al. 1992; Schweizer et al. 1995). Detection of PFV proviral DNA by PCR was reported in lymphocytes of 19 of 29 patients with Graves disease. PFV sequences were also detectable in PBL DNA by Southern blot hybridization (Lagaye et al. 1992). However, all patient sera were negative for antibodies to PFV. Because FV infection is generally associated with a strong humoral immune response, the lack of FV specific antibodies in these patients is uncharacteristic of genuine FV infection. Evidence of PFV antigen in biopsy sections was also reported in different studies of patients with thyroiditis by using immunofluorescence assays with polyclonal rabbit antisera against PFV antigens (Wick et al. 1992, 1993). Several studies of different patient cohorts failed to find molecular or serologic evidence of PFV infection in patients with Graves disease (Mahnke et al. 1992; Schweizer et al. 1994, 1995; Heneine et al. 1995; Yanagawa et al. 1995). A report on patients with sudden deafness and Ménière disease showing evidence of seropositivity to PFV by immunofluorescence testing (Pyykkö et al. 1993) was not confirmed by others (Simonsen et al. 1994). The antibody reactivity was later recognized as cross-reactivity with a cellular antigen, and the report on the association with FV infection was retracted (Pyykkö et al. 1994).

The finding of antibody reactivity to PFV in patients with amyotrophic lateral sclerosis (ALS) was of particular significance, because data from these studies led to attempts to treat some patients with ALS with antiretroviral drugs (Zidovudine). Serologic screening in these studies was made by two ELISAs using different recombinant PFV proteins, and confirmation was done by Western blot with antigen of unexplained origin. Reactivity was found to be higher in ALS patients (47%) than in healthy controls (21%) (Westarp et al. 1992, 1993a–d, 1994). However, an independent study could not find any evidence of PFV infection in ALS patients (Rösener et al. 1996). Isolated reports of PFV DNA amplification in patients with familial Mediterranean fever or Graves disease patients as well as healthy persons from Korea have also been published (Tamura et al. 1995; Lee et al. 1998). However, additional evidence of infection including detection of FV-specific antibodies was missing in these studies. The absence of PFV infection is also highlighted in several diseases including multiple sclerosis (Svenningson et al., 1992), chronic fatigue syndrome (Gow et al. 1992; Baum 1992; Gunn et al. 1993; Heneine et al. 1994), various cancers, as well as idiopathic CD4+-T-cell lymphocytopenia (Schweizer et al. 1995). A case report of a FV infection in one of eight patients with myasthenia gravis described serologic and molecular evidence of FV infection (Saïb et al. 1994). However, no independent confirmation of these results is available.

5
Identification of SFV Infection in Humans

In contrast to the absence of evidence for FV infection seen in humans, SFVs have been identified in almost all NHP species examined of both Old World and New World origin, and prevalence rates exceeding 70% have been seen in adult captive NHP colonies (Hooks 1975; Neumann-Haefelin et al. 1993; Schweizer et al. 1995). The concordance found between the phylogeny of SFV and their host NHP species suggests a long-standing coexistence and coevolution (Herchenroder et al. 1994; Bieniascz et al. 1995; Schweizer 1995; Heneine et al. 1998). The identification of two isolated occupationally acquired infections with SFV of chimpanzee and African green monkey (AGM) origin suggested that occupational contact with NHP or SFV is associated with risks of SFV transmission to humans (Schweizer et al. 1995; Schweizer et al. 1997). A study conducted by the US Centers for Disease Control and Prevention (CDC) on workers occupationally exposed to NHP, their body fluids, or viruses provided information on the prevalence of SFV infection in this population and evaluated clinical outcome and risks of secondary transmission associated with such infections (Heneine et al. 1998). Of 231 persons enrolled from 13 institutions in North America, 4 (1.8%) were found to be SFV infected by both Western blot and PCR testing (Fig. 1). Virus was also isolated from PBL from two of the four persons. Analysis of the integrase sequence indicated that the infecting SFV originated from baboons in three cases and AGM in one case (Fig. 2). All four workers were males and all confirmed frequent exposure to body fluids of the implicated species, and in some but not all instances injuries were associated with these NHP species (Heneine et al. 1998). Testing of archived sera indicated long-standing infections with durations of documented seropositivity ranging between 5 and 23 years. Although this study clearly demonstrates a substantial rate of SFV infection in persons occupationally exposed to NHP, biases that favor voluntary enrollment of workers with increased likelihood of infection may likely overestimate the prevalence of infection in this population. This study at CDC is ongoing, and preliminary data indicate that additional SFV-infected persons have been identified.

Evidence of SFV infection was also found in an unlinked serosurvey of 322 North American zoo workers (Sandstrom et al. 2000). This study identified four Western blot-positive samples. The four reactive sera were subsequently tested against antigens from SFV-6 (chimpanzee), SFV-3 (African green monkey), and SFV-2 (macaque) separately. Patterns of seroreactivity suggested that all four were infected with chimpanzee-like SFV. All 4 samples were from the 133 workers whose jobs involved potential contact with NHP. None of 189 workers whose jobs did not involve potential contact with NHP were seroreactive.

6 Clinical Outcome and Human-to-Human Transmissibility

Although SFV infections in naturally infected NHP are known to be nonpathogenic, fewer data are available about the pathogenicity of SFV infections in humans. The findings reported to date on six SFV-infected persons indicate absence of disease despite the evidence of long-standing SFV seropositivity documented in several workers (Schweizer et al. 1995, 1997; Heneine et al. 1998). The lack of disease suggests that transmission of SFV to humans is not associated with evident change in pathogenicity. However, these findings are limited to a small number of infected persons and cannot exclude the occurrence of disease after long latency periods or the possible association of disease with SFV clades of

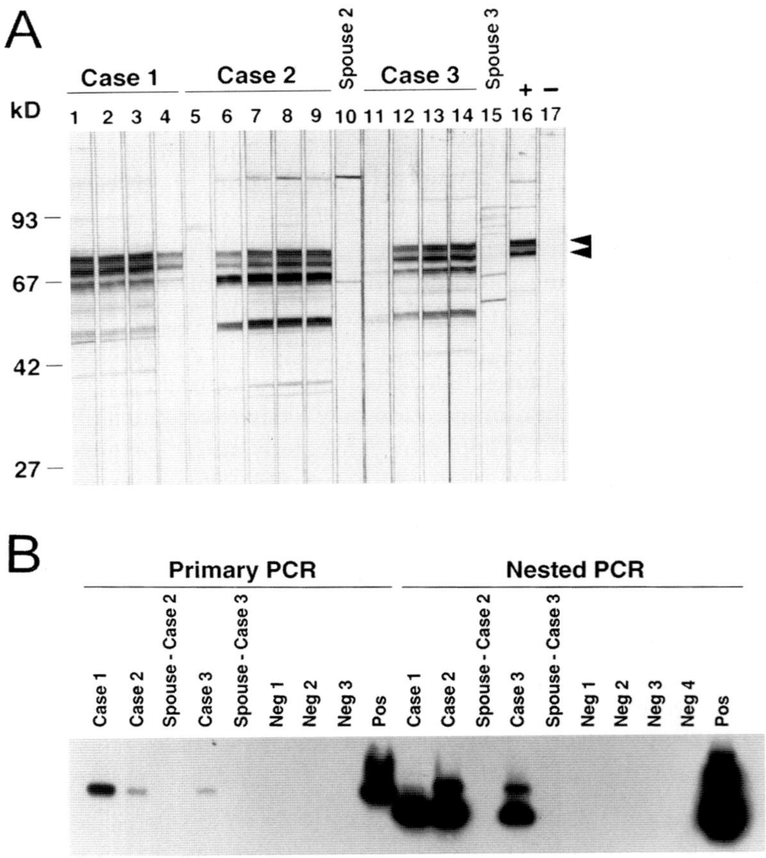

Fig. 1A, B. Laboratory diagnosis of SFV infection in case workers occupationally exposed to nonhuman primates. **A** Western blot reactivity to SFVagm antigen. *Arrows*, diagnostic p68 and p71 bands. Lanes *1–4*, sera collected in 1995, 1996, Jan. 1997, and Feb. 1997; lanes *5–9*, sera collected in 1967, 1978, 1984, 1990, and 1996; lane *10*, serum from spouse of case 2 collected in 1996; lanes *11–14*, sera collected in 1988, 1994, 1996, and 1997; lane *15*, serum from spouse of case 3 collected in 1996; lanes *16* and *17*, SFV-infected and SFV-uninfected African green monkeys, respectively. **B** PCR analysis of peripheral blood lymphocytes of a 153 bp FV *pol* sequence. Southern blot hybridization results of PCR products from case workers 1, 2, 3 and two spouses. *Neg 1*, SFV-uninfected cell control; *Neg 2, 3, 4*, water; *Pos*, SFV-infected cell control

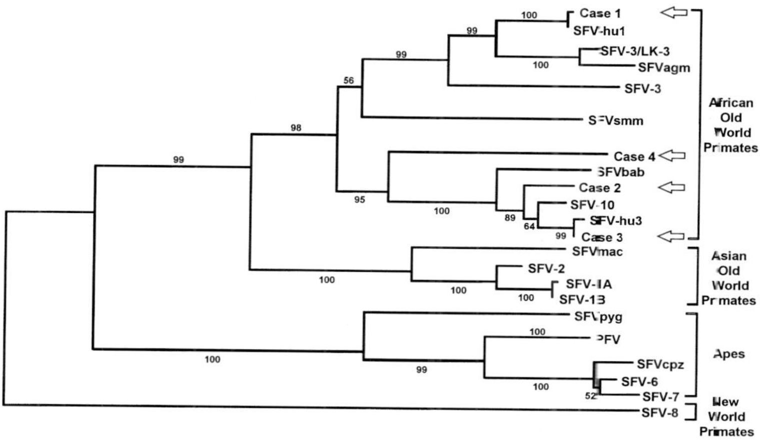

Fig. 2. Nonhuman primate (*NHP*) species origin of SFV infections in occupationally exposed workers. Phylogenetic analysis by neighbor joining of FV *pol* sequences derived from peripheral blood lymphocytes of four cases and from SFVhu1 and SFVhu3, two SFV isolates obtained from case 1 and case 3, respectively. SFV sequences of different NHP species shown are SFVagm, SFVsmm, SFVbab, SFVmac, and SFVpyg, obtained from African green monkey, sooty mangabey, baboon, and pygmy chimpanzee, respectively, SFVcpz and SFV-6 and SFV-7, from chimpanzees, and SFV-8 from spider monkey (a New World monkey species). *Numbers* shown at branch nodes represent bootstrap percent values

different NHP species origin. The risk for secondary transmission has also been examined in a few SFV-infected persons. The ability of SFV to spread from infected persons to their contacts has important public health implications and is of concern because of the ease with which SFV transmits among NHP. The data reported to date show no evidence of male-to-female transmission in wives of four SFV-infected workers (Heneine et al. 1998; Schweizer et al. 1997). Although limited, these findings suggest that transmission of SFV among humans by sexual or less intimate contact may not occur easily. The absence of disease and secondary transmission suggest that SFVs may represent benign end point infections. However, several SFV-infected workers identified in the CDC study have reported a history of blood donation, raising questions about SFV spread by transfusion of blood or blood products. A study at CDC of recipients of blood products from an SFV-infected donor showed no evidence of SFV in two recipients of red cells, one recipient of filtered red cells, and one recipient of platelets (Boneva et al. 2002). All SFV-infected persons should be counseled to refrain from blood and tissue donation until risks of transmission by blood are well defined.

7 Genetic Stability of SFV and Adaptive Changes in the Human Host

Limited information is available on the mechanism of persistence of SFV and the genetic stability during chronic infection. In a longitudinal study of captive African green monkeys it was found that the *env* gene of SFV demonstrates remarkable intrahost stability, likely reflecting both the low replication rate of SFV as well as the absence of positive selective pressures in the natural host (Schweizer et al. 1999). However, SFV demonstrates significant diversity within the U3 region of the LTR and the 3' accessory genes among different species of simians, suggesting that these genetic regions may be prone to adaptive changes during cross-species infection. Data on changes in these sequences after human infection are very limited. A study of one SFV-infected person shows a high sequence conservation indicating little host-specific adaptation in these sequences (Callahan et al. 1999). The genetic stability of SFV in humans may be consistent with an apparent lack of pathogenicity and will likely be an important consideration in the design of FV vectors for human gene therapy.

8 Prevalence of Feline Foamy Virus Infection Among Humans

The observation that SFV appears to be readily transmissible to humans raises concerns that human contact with other domestic species such as cats, cattle, and horses, all of which are known to harbor endemic FV infection (Loh 1992; Tobaly-Tapiero et al. 2000), may be accompanied with the risk of transmission of FV. Although bovine or equine FV zoonosis in the general population or among occupationally exposed persons has not been examined, there have been two recently published studies addressing the risk of cross-species infection with feline FV (FFV) (Butera et al. 2000; Winkler et al. 1997). FFV appears to be readily transmitted among domestic cats via infectious saliva either through biting, communal eating, or grooming (Winkler et al. 1999), resulting in a prevalence of infection of 30%–90% (Winkler et al. 1998). Similar to SFV, FFV can be easily cultured in human cell lines, supporting the susceptibility of human cells to infection with this FV (Butera et al. 2000).

Human exposure to FFV in the United States might be expected to be extensive when one considers the widespread use of domestic felines as pets. However, studies looking for evidence of FFV transmission to occupationally exposed humans suggest that the risk of FV zoonosis from species outside of Old World primates may be remote. A recent large cross-sectional study of 203 North American feline veterinary practitioners found no serologic evidence of FFV infection (Butera et al. 2000). Individuals within this group reported an average of 17.3 years of occupational exposure to domestic cats and experienced on average three cat bite wounds per year, five severe cat scratch wounds per year, and four needle injuries per year. A limited number of respondents also reported additional types of high-risk contacts including exposure to feline blood, urine, or abscess fluid on broken skin and puncture wounds with bone marrow aspirates or fixation pins. However, despite these high-risk exposures, no evidence of FFV was detected in any of the subjects. These findings agree with those of an earlier study reporting a lack of FFV zoonosis among Australian veterinarians, although exposure to felines and FFV were not clearly defined in this study (Winkler et al. 1997).

The reasons for this apparent species restriction of FV zoonosis are not clear but likely reflect an innate human resistance to retroviral infections arising from mammal species other than Old World primates (Brown et al. 1998). Owing to a frame shift mutation in the $\alpha 1$–3 galactose transferase gene, humans and Old World primates are unique among mammals in having no $\alpha 1$–3 galactose (α-gal) residues on cell surface glycoproteins and glycolipids (Larsen et al. 1990). Anti-α-gal antibodies present in normal human serum have been shown to effectively neutralize retroviruses (including FV) from other mammalian species whose envelope structures incorporate host cell specific $\alpha 1$–3 gal motifs (Rother et al. 1995).

9 Conclusions

Although virtually all NHP species investigated thus far including prosimians, New World and Old World monkeys, and ape species all harbor distinct and species-specific clades of SFV, evidence supporting the existence of a human-specific FV is not yet available. Instead, all FV

Table 1. Unanswered questions in the natural history of SFV infections in humans

1. Disease association not fully defined because numbers of known SFV-infected persons are low
2. Impact of route of infection, dose, and SFV strain/clade on disease development?
3. Do risks for disease development increase in immunosuppressed hosts?
4. Role of immune responses to SFV in protecting against transmission and disease?
5. What are the risks of human-to-human spread by sexual contact, blood donation, from mother-to-child, or by other contacts?
6. What is the tissue distribution of SFV in humans and its role in transmissibility?
7. Would successful human-to-human transmission of SFV enhance transmissibility or pathogenicity?

infections identified in humans are of zoonotic origin and are seen in persons occupationally exposed to NHP. The reasons why humans, despite common evolution and long periods of cohabitation with NHP populations, are not endemically infected with a distinct FV are not understood. Nevertheless, the documented introduction of SFV into humans through contact with NHP raises several public health questions regarding disease outcomes and potential for human-to-human transmissibility. These questions about the natural history of SFV in humans are summarized in Table 1. Available data from a limited number of SFV-infected humans suggest that these infections are nonpathogenic and are not easily transmissible. However, additional studies of a larger number of infected persons are necessary to confirm the benign nature of SFV in humans and to address several unanswered questions. For instance, it is not known whether the susceptibility of humans to SFV infection and disease will be clade dependent and whether distinct SFV clades of different NHP species origin will pose risks to infected humans. Because all present data are based on primary cross-species infections, it is not known whether certain secondary exposures like transfusion of infected donated blood carry higher risks for transmission and whether successive blood-borne transmissions among humans may enhance opportunities for SFV to adapt and become more transmissible or pathogenic. Therefore, a better understanding of the natural history of SFV infections in humans is necessary. This information will be important both for facilitating protection of occupationally exposed workers and for evaluating the safety of SFV-based vectors currently under development for vaccine and therapeutic gene delivery.

References

Ali M, Taylor GP, Pitman RJ, Parker D, Rethwilm A, Cheingsong-Popov R, Weber JN, Bieniasz PD, Bradley J, McClure MO (1996) No evidence of antibody to human foamy virus in widespread human populations. AIDS Res Hum Retrovir 12:1473–1483

Achong BG, Mansell PWA, Epstein MA, Clifford P (1971) An unusual virus in cultures from a human nasopharyngeal carcinoma. J Natl Cancer Inst 46:299–302

Achong BG, Epstein MA (1978) Preliminary seroepidemiological studies on the human syncytial virus. J Gen Virol 40:175–181

Achong BG, Epstein MA (1983) Naturally occurring antibodies to the human syncytial virus in West Africa. J Med Virol 11:53–57

Baum KF (1992) Absence of antibody to human spumaretrovirus in patients with chronic fatigue syndrome. Clin Infect Dis 14:623–624

Bieniasz PD, Rethwilm A, Pitman R, Daniel MD, Chrystie I, McClure MO (1995) A comparative study of higher primate foamy viruses, including a new virus from a gorilla. Virology 207:217–228

Boneva RS, Grindon AJ, Orton SL, Switzer WM, Shamugam V, Hussain AI, Bhullor VB, Chamberland ME, Heneine W, Folks TM, Chapman LE (2002) Simian foamy virus infection in a blood donor. Transfusion 42:886–891

Brown J, Matthews AL, Sandstrom PA, Chapman LE (1998) Xenotransplantation and the risk of retroviral zoonosis. Trends Microbiol 6:411–415

Brown P, Nemo G, Gajdusek DC (1978) Human foamy virus: further characterization, seroepidemiology, and relationship to chimpanzee foamy viruses. J Infect Dis 137:421–427

Butera ST, Brown J, Callahan ME, Owen SM, Matthews AL, Weigner DD, Chapman LE, Sandstrom PA (2000) Survey of veterinary conference attendees for evidence of zoonotic infection by feline retroviruses. J Am Vet Assoc 217:1475–1479

Callahan ME, Switzer W, Matthews AL, Roberts B, Heneine W, Folks TM, Sandstrom PA (1999) Persistent zoonotic infection of a human with simian foamy virus in the absence of an Intact orf-2 accessory gene. J Virol 73:9619–9624

Cameron KR, Birchall SM, Moses MA (1978) Isolation of foamy virus from patient with dialysis encephalopathy. Lancet II:796

Cordonnier A, Casella JF, Heidmann T (1995) Isolation of novel human endogenous retrovirus-like elements with foamy virus-related pol sequence. J Virol 69:5890–5897

Debons-Guillemin MC, Valla J, Gazeau J, Wybier-Franqui J, Giron ML, Toubert ME, Canivet M, Périès J (1992) No evidence of spumaretrovirus infection markers in 19 cases of de Quervain's thyroiditis. AIDS Res Hum Retrovir 8:1547

Gow JW, Simpson K, Schliephake A, Behan WM, Morrison LJ, Cavanagh H, Rethwilm A, Behan PO (1992) Search for retrovirus in the chronic fatigue syndrome. J Clin Pathol 45:1058–1061

Gunn WJ, Komaroff AL, Connell DB, Levine SM, Cheney PR (1993) Inability of retroviral tests to identify persons with chronic fatigue syndrome. MMWR Morb Mortal Wkly Rep 42:189–190

Hahn H, Baunach G, Bräutigam S, Mergia A, Neumann-Haefelin D, Daniel MD, McClure M, Rethwilm A (1994) Reactivity of primate sera to foamy virus Gag and Bet proteins. J Gen Virol 75:2635–2644

Heneine W, Woods TC, Sinha SD, Khan AS, Chapman LE, Schonberger LB, Folks TM (1994) Lack of evidence for infection with known human and animal retroviruses in patients with chronic fatigue syndrome. Clin Infect Dis 18 Suppl 1:121–125

Heneine W, Musey VC, Sinha SD, Landay A, Northrup G, Khabbaz R, Kaplan JE (1995) Absence of evidence for human spumaretrovirus sequences in patients with Graves' disease. J Acquir Immun Defic Syndr Hum Retrovir 9:99–101

Heneine W, Switzer WM, Sandstrom P, Brown J, Shanmugam V, Schable CA, Khan A S; Lerche NW, Schweizer M, Neumann-Haefelin D, Chapman LE, Folks TM (1998) Identification of a human population infected with simian foamy virus. Nat Med 4:403–407

Herchenröder O, Renne R, Loncar D, Cobb EK, Murthy KK, Schneider J, Mergia A, Luciw PA (1994) Isolation, cloning, and sequencing of simian foamy viruses from chimpanzees (SFVcpz): High homology to human foamy virus (HFV). Virology 201:187–199

Hooks JJ, Gibbs CJ (1975) The foamy viruses. Bacteriol Rev 39:169–185

Hunsmann G, Flügel RM, Walder R (1990) Retroviral antibodies in Indians. Nature 344:120

Lagaye S, Vexiau P, Morozov V, Guénebaut-Claudet V, Tobaly-Tapiero J, Canivet M, Cathelineau G, Périès J, Emanoil-Ravier R (1992) Human spumaretrovirus-related sequences in the DNA of leukocytes from patients with Graves disease. Proc Natl Acad Sci USA 89:10070–10074

Larsen RD, Rivera-Marrero CA, Ernst LK, Cummings RD, Lowe JB (1990) Frameshift and nonsense mutations in a human genomic sequence homologous to a murine UDP-Gal:beta-D-Gal(1,4)-D-GlcNAcalpha(1,3) -galactosyltransferase cDNA. J Biol Chem 265:7055-61

Lee H, Kim S, Kang M, Kim W, Cho B J (1998) Prevalence of human foamy virus-related sequences in the Korean population. Biomed Sci 5:267–73

Loh PC, Matsuura F, Mizumoto C (1980) Seroepidemiology of human syncytial virus: antibody prevalence in the Pacific. Intervirology 13:87–90

Loh PC. Spumaviruses. In: Levy JA (ed) The Retroviridae, Plenum Press, New York

Mahnke C, Kashaiya P, Rössler J, Bannert H, Levin A, Blattner WA, Dietrich M, Luande J, Löchelt M, Friedman-Kien AE, Komaroff AL, Loh PC, Westarp ME, Flügel RM (1992) Human spumavirus antibodies in sera from African patients. Arch Virol 123:243–253

McClure MO, Erlwein O (1995) Foamy viruses—pathogenic or therapeutic potential. Rev Med Virol 5:229–237

Muller HK, Ball G, Epstein MA, Achong BG, Lenoir G, Levin A (1980) The prevalence of naturally occurring antibodies to human syncytial virus in East African populations. J Gen Virol 47:399–406

Nemo GJ, Brown PW, Gibbs CJ Jr, Gajdusek DC (1978) Antigenic relationship of human foamy virus to the simian foamy viruses. Infect Immun 20:69–72

Neumann-Haefelin D, Fleps U, Renne R, Schweizer M (1993) Foamy viruses. Intervirology 35:196–207

Pyykkö I, Vesanen M, Asikainen K, Koskiniemi M, Airaksinen L, Vaheri A (1993) Human spumaretrovirus in the etiology of sudden hearing loss. Acta Otolaryngol 113:109–112

Pyykkö I, Vesanen M, Asikainen K, Koskiniemi M, Airaksinen L, Vaheri A (1994) Human spumaretrovirus in the etiology of sudden hearing loss. Acta Otolaryngol 114:224

Rösener M, Hahn H, Kranz M, Heeney J, Rethwilm A (1996) Absence of serological evidence for foamy virus infection in patients with amyotrophic lateral sclerosis. J Med Virol 48:222–226.

Rother RP, Fodor WL, Springhorn JP, Birks CW, Setter E, Sandrin MS, Squinto SP, Rollins SA (1995) A novel mechanism of retrovirus inactivation in human serum mediated by anti-alpha-galactosyl natural antibody. J Exp Med 182:1345-1355

Saïb A, Périès J, de Thé H (1993) A defective human foamy provirus generated by pregenome splicing. EMBO J 12:4439–4444

Saïb A, Canivet M, Giron ML, Bolgert F, Valla J, Lagaye S, Périès J, de Thé H (1994) Human foamy virus infection in myasthenia gravis. Lancet 343:666

Sandstrom PA, Phan K, Switzer W, Chapman L, Heneine W, Folks TM (2000) Simian foamy virus infection among zoo keepers. Lancet 355:551–552

Schweizer M, Turek R, Reinhardt M, Neumann-Haefelin D (1994) Absence of foamy virus DNA in Graves' disease. AIDS Res Hum Retrovir 10:601–605

Schweizer M, Turek R, Hahn H, Schliephake A, Netzer KO, Eder G, Reinhardt M, Rethwilm A, Neumann-Haefelin D (1995) Markers of foamy virus (FV) infections in monkeys, apes, and accidentally infected humans: Appropriate testing fails to confirm suspected FV prevalence in man. AIDS Res Hum Retrovir 11:161–170

Schweizer M, Neumann-Haefelin D (1995) Phylogenetic analysis of primate foamy viruses by comparison of pol sequences. Virology 207:577–582

Schweizer M, Falcone V, Gänge J, Turek R, Neumann-Haefelin D (1997) Simian foamy virus isolated from an accidentally infected human individual. J Virol 71:4821–4824

Schweizer M, Schleer H, Pietrek M, Liegibel J, Falcone V, Neuman-Haefelin D (1999) Genetic Stability of foamy viruses: Long-term study in an African green monkey population. J Virol 73:9256–9265

Simonsen L, Heneine W, Sinha SD, Arenberg IK (1994) Absence of evidence for infection with the human spuma retrovirus in an outbreak of Meniere-like vertiginous illness in Wyoming, USA. Acta Oto-Laryngologica 114:223–224

Stancek D, Gressnerová M (1974) A viral agent isolated from a patient with subacute de Quervain type thyroiditis. Acta Virol 18:365

Stancek D, Stanceková-Gressnerova M, Janotka M, Hnilica P, Cravec D (1975) Isolation and some serological and epidemiological data on the viruses recovered from patients with subacute thyroiditis de Quervain. Med Microbiol Immunol 161:133–144

Stancekova M, Stancek D, Ciampor F, Mucha V, Hnilica P (1976) Morphological, cytological and biological observations on viruses isolated from patients with subacute thyroiditis de Quervain. Acta Virol 20:183–188

Svenningsson A, Lycke J, Svennerholm B, Gronowitz S, Andersen O (1992) No evidence for spumavirus or oncovirus infection in relapsing-remitting multiple sclerosis. Ann Neurol 32:711–714

Tamura N, Kira S (1995) Human foamy virus and familial Mediterranean fever in Japan. JAMA 274:1509

Tobaly-Tapiero J, Bittoun P, Neves M, Guillemin MC, Lecellier CH, Puvion-Dutilleul F, Gicquel B, Zientara S, Giron ML, de The H, Saïb A (2000) Isolation and characterization of an equine foamy virus. J Virol 74:4064-4073

Von Laer D, Neumann-Haefelin D, Heeney JL, Schweizer M (1996) Lymphocytes are the major reservoir for foamy viruses in peripheral blood. Virology 221:240–244

Wallen WC, Sever JL, McFarlin DE, McFarland HF, Traub RG, Rentier B, Greenstein JI, and Moore PM (1979) Attempt to isolate infectious agent from bone-marrow of patients with multiple sclerosis. Lancet II:414–415

Weiss RA (1988) Foamy retroviruses: A virus in search of a disease. Nature 333:497–498

Werner J, Gelderblom H (1979) Isolation of foamy virus from patients with de Quervain thyroiditis. Lancet II:258–259

Westarp ME, Kornhuber HH, Rössler J, Flügel RM (1992) Human spuma retrovirus antibodies in amyotrophic lateral sclerosis. Neurol Psychiatr Brain Res 1:1–4

Westarp ME, Bartmann P, Hoff-Jorgensen R, Clausen J, Rasmussen H, Kornhuber HH (1993a) Amyotrophe Lateralsklerose–Hinweise fur eine erhöhte antiretrovirale Seroreaktivitat ohne augenfallige Epidemiologie. Nervenarzt 64:384–389

Westarp ME, Fuchs D, Bartmann P, Hoff-Jorgensen R, Clausen J, Wachter H, Kornhuber HH (1993b) Amyotrophic lateral sclerosis an enigmatic disease with B-cellular and anti-retroviral immune responses. Eur J Med 2:327–332

Westarp ME, Westphal KP, Clausen J, Rasmussen HB, Hoff-Jorgensen R, Fohring B, Kornhuber HH (1993c) Retroviral interference with neuronotrophic signaling in human motor neuron disease? Clin Physiol Biochem 10:1–7

Westarp ME, Bartmann P, Rossler J, Geiger E, Westphal KP, Schreiber H, Fuchs D, Westarp MP, Kornhuber HH (1993d) Antiretroviral therapy in sporadic adult amyotrophic lateral sclerosis. Neuroreport 4:819–822

Westarp ME, Foring B, Rasmussen H, Schraff S, Mertens T, Kornhuber HH (1994) Retroviral synthetic peptide serum antibodies in human sporadic amyotrophic lateral sclerosis. Peptides 15:207–214

Wick G, Grubeck-Loebenstein B, Trieb K, Kalischnig G, Aguzzi A (1992) Human foamy virus antigens in thyroid tissue of Graves' disease patients. Int Arch Allergy Immunol 99:153–156

Wick G, Trieb K, Aguzzi A, Recheis H, Anderl H, Grubeck-Loebenstein B (1993) Possible role of human foamy virus in Graves' disease. Intervirology 35:101-107

Winkler IG, Löchelt M, Levesque JP, Bodem J, Flügel RM, Flowers RLP (1997) A rapid streptavidin-capture ELISA specific for the detection of antibodies to feline foamy virus. J Immunol Meth 207:69–77

Winkler IG, Flügel RM, Löchelt M, Flower RLP (1998) Detection and molecular characterization of feline foamy virus serotypes in naturally infected cats. Virology 247:144–151

Winkler IG, Löchelt M, Flower RLP (1999) Epidemiology of feline foamy virus and feline immunodeficiency virus infections in domestic and feral cats: a seroepidemiological study. J Clin Micro 37:2848–2851

Yanagawa T, Ito K, Kaplan EL, Ishikawa N, DeGroot LJ (1995) Absence of association between human spumaretrovirus and Graves' disease. Thyroid 5:379–382

Chapter 9

Non-primate Foamy Viruses

A. Saïb

Abstract Foamy viruses (PFVs), also called spumaviruses, are complex retroviruses inducing a characteristic cytopathic effect in cell culture, leading rapidly to cell lysis. These viruses have been isolated mostly in non-human primates, but three non primate PFVs were characterized, namely the bovine foamy virus, the feline foamy virus and more recently the equine foamy virus. In their hosts, PFVs seem to be apathogenic, mirroring an efficient control of virus replication in vivo. Comparing the biology of the different virus isolates will certainly help to unravel the biology of these retroviruses.

A. Saïb (✉)
CNRS UPR9051, Hôpital Saint-Louis,
1 Avenue Claude Vellefaux, 75475 Paris Cedex 10, France
E-mail: alisaib@infobiogen.fr

1 Introduction

Foamy viruses (PFVs) harbour several features which distinguish them from other retroviruses. The most evident of them lies in their name. In vitro, PFVs induce the formation of multinucleated giant cells, resulting from the fusion of adjacent infected cells, which present multiple vacuoles giving the monolayer culture a foamy aspect (Fig. 1; Saïb et al. 1995). The observation of this specific cytopathic effect, together with electron microscopy studies, helped investigators in their attempt to isolate new PFVs. Since 1954, the year when Enders and Peebles isolated the first PFV from rhesus monkey kidney cells (Enders and Peebles 1954), these viruses have been described in different animal species, mostly in non-human primates (Flügel 1991). Similarly, non-primate PFVs have been described, isolated and further characterised. Three non-primate PFVs have been well studied: the bovine foamy virus (BFV), the feline foamy virus (FFV) and, more recently, the equine foamy virus (EFV). Note that other non-primate PFVs have been isolated or simply described in sea lions, goats, sheep and finally hamsters on the basis of cross-antigenicity with known foamy viruses, appearance of specific CPE or electron microscopy analysis (Kennedy-Stoskopf et al. 1986; Flanagan 1992; Hruska and Takemoto 1975, reviewed in Lecellier and Saïb 2000).

In this chapter, I present these isolates and compare them to their primate counterparts.

2 History and Epidemiology of Non-primate PFVs

2.1 The Bovine Foamy Virus

Since the 1960s, epidemiologic observations have suggested the viral aetiology of lymphosarcoma in cattle (Dutcher et al. 1967). In that sense, several publications have described the presence of viruslike particles in tissues from affected animals. In 1969, a syncytium-inducing agent was isolated from lymph node, spleen and milk sediment in 4 of 15 cattle with lymphosarcoma (Malmquist et al. 1969). This viral agent was shown to induce syncytia formation in bovine embryonic spleen cul-

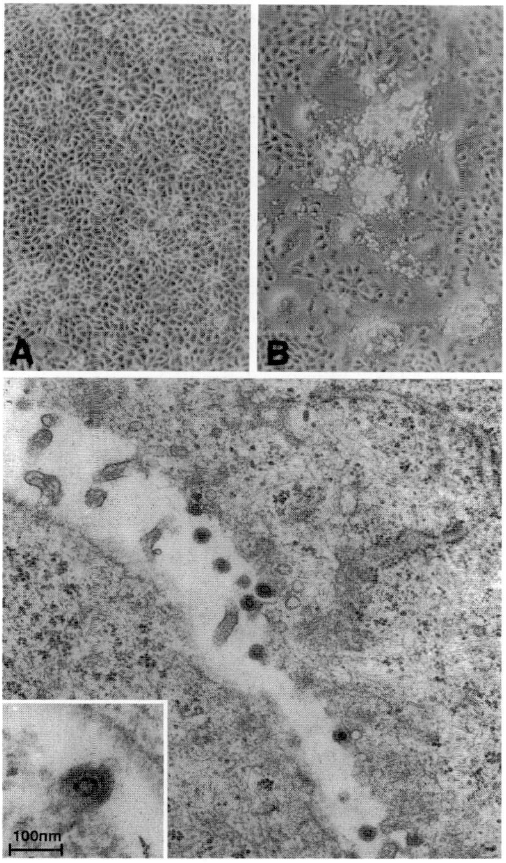

Fig. 1A, B. Characteristics of foamy virus infection. **A** Uninfected BHK-21 cells. **B** EFV-infected cells presenting multiple vacuoles. PFVs induce the formation of giant multinucleated cells presenting vacuoles, giving the cell culture a foamy aspect. *Bottom*: Virions as visualised by electron microscopy. Virus particles harbour numerous spikes of 10–15 nm and mainly bud from the plasma membrane in the case of EFV, whereas other PFVs bud mostly into intracytoplasmic vesicles (see text). *Bar*, 100 nm

tures, leading to rapid cell lysis. Electron microscopy studies further revealed the presence of an enveloped virus harbouring protuberant projections (14–18 μm), similarly as reported for primate PFVs. However, the presence of this agent in a large number of healthy cattle did not favour a direct link between this virus and lymphosarcoma (Johnston

1974, 1988). Indeed, serological tests (immunodiffusion, Western blot, indirect immunofluorescence assays) and recent molecular analysis (enzymatic amplification) demonstrated that between 30% and 45% of cattle were infected with the bovine foamy virus (BFV, initially called bovine syncytial virus or bovine spumavirus) (Kertayadnya et al. 1988; Jacobs et al. 1995; Pamba et al. 1999).

The molecular structure of BFV was published much later by Renshaw et al., who described the genetic map and the transcriptional status of the bovine isolate, showing that this provirus of 12 kb harbours the same features as its primate counterpart PFV: two LTRs (long terminal repeats) at the extremities of the viral genome, an internal single strand gap on the plus strand DNA and a 18-bp PBS (primer binding site, complementary to the 3' end of the cellular $tRNAlys_{1,2}$ specific to all PFVs), for example (Renshaw and Casey 1994a,b; Renshaw et al. 1991).

2.2
The Feline Foamy Virus

Also in 1969, another PFV was isolated from domestic, normal and diseased, cats based on the appearance of foamy cytopathic effects after in vitro seeding of fresh tissue samples (Riggs et al. 1969; Fabricant et al. 1969). Morphological, biological and biochemical analyses demonstrated that FFV (initially called FSFV or FeSFV for feline syncitium-forming virus) belonged to the spumavirus subfamily of *Retroviridae*. Similar to BFV, FFV is widespread among wild and domestic cats as revealed by immunological surveys which showed that from 31% to over 70% of cats are FFV-seropositive, depending on their age or geographic origin (Winkler et al. 1997b, 1998,. 1999; Daniels et al. 1999). Although no pathology has been linked to FFV, the recurrent co-isolation of FFV in FIV-infected cats has raised the possibility of a role of cofactor in the onset of feline acquired immunodeficiency. However, it has been demonstrated that FFV does not facilitate progression of FIV-induced disease, although both viruses seem to be transmitted via a common mode (Winkler et al. 1999; Zenger et al. 1993).

Although the provirus size and structure of the viral DNA were biochemically investigated in the late 1970s (Chiswell and Pringle 1977,1979a,b), its molecular cloning as well as its expression pattern have only been recently reported (Bodem et al. 1996; Helps and Harbour 1997; Winkler et al. 1997a). This analysis led to the identification of two distinct serotypes, the major difference being confined to the surface SU Env glycoprotein.

More recently, replication-competent FFV vectors were constructed and functionally tested. Viruses derived from the cloned FFV DNA replicated and persisted in experimentally infected cats, similar to the FFV isolate FUV. An FFV vector, under the transcriptional control of the strong human cytomegalovirus immediate-early promoter, and partially deleted in the non-coding area of the U3 region, was used to transduce the GFP, expressed either from an internal ribosomal entry site (IRES) or as C-terminal fusion protein linked to Bet. Whereas the genetic stability of the IRES-GFP construct was comparably low, the Bet-GFP fusion protein was detectable on serial cell free vector passages. However, genetic rearrangements also occurred, leading to the concomitant loss of marker gene expression (A. Schwantes and M. Löchelt, personal communication).

2.3 The Equine Foamy Virus

Lentiviruses were isolated in primates, cattle, cats, sheep, goats and horses. Strikingly, the distribution of PFVs mirrors that of lentiviruses, except in the case of horses for which no data were available. Thus the presence of a foamy virus was recently investigated in domestic horses, and this has led to the isolation of a new PFV isolate, which has been called EFV. After co-culture of fresh horse blood peripheral cells with permissive human U373MG cells, a foamlike cytopathic effect was rapidly observed and PFV virions were clearly present in these cells as demonstrated by electron microscopy. Finally, a18-bp PBS sequence used as a probe enabled us to clone the entire viral genome, demonstrating that this virus belongs to the *Spumavirinae* subfamily (Tobaly-Tapiero et al. 2000). Recently, experimental infections of rabbits led to the production of polyclonal antibodies directed against EFV polypeptides, allowing us to study these viral proteins at the biochemical level (Lecellier et al. 2002b).

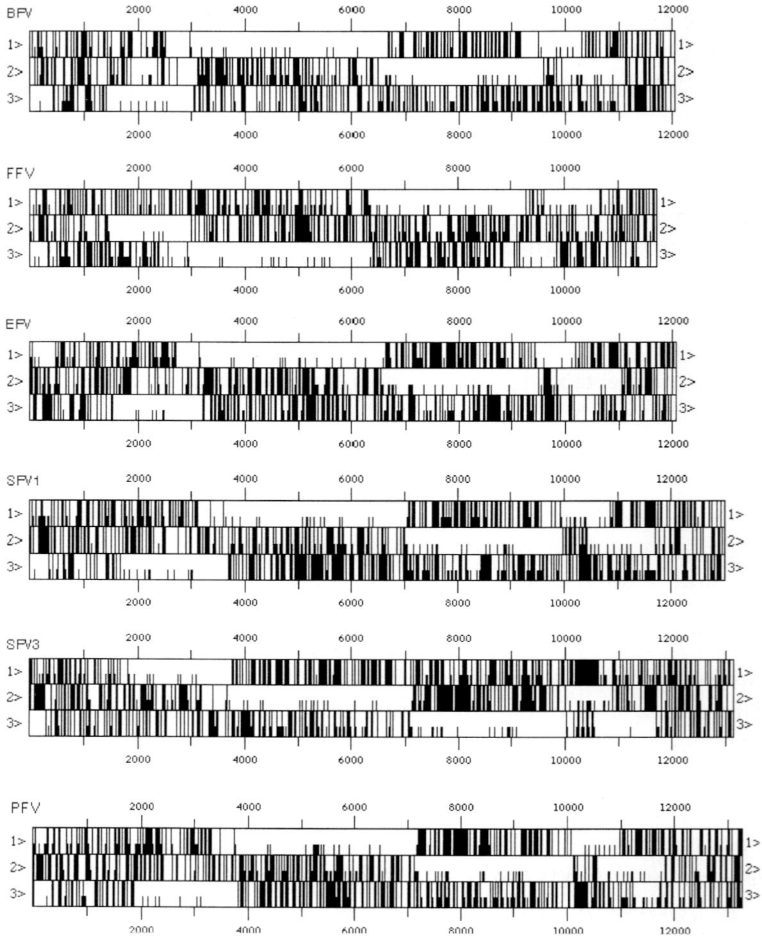

Fig. 2. Open reading frames of primate and non-primate PFVs deducted from computer analysis. Note the presence of two regulatory ORFs at the 3' end of all PFVs, except PFV, which harbours a third ORF

3
Molecular Characteristics of Non-primate PFVs: Similarities and Differences with Primate PFVs

The genomes of non-primate PFVs closely resemble those of the primate isolates: similar in size, they harbour the three structural and enzymatic genes, *gag*, *pol* and *env*, and regulatory genes at the 3' end of the provirus (Fig. 2). Computer analysis and analogy studies with other sequenced PFVs were used to determine predicted sizes and conserved features between primate and non-primate PFVs.

3.1 The *gag* Gene

Although non-primate PFV Gag polypeptides are smaller than their primate counterparts (between 88 and 145 residues), the overall structure appears identical as revealed by their hydrophilicity plots (Helps and Harbour 1997). One specific feature associated to this retroviral subfamily is the absence of the classical tripartite cleavage into matrix, capsid and nucleocapsid mature products (Pfrepper et al. 1999). Indeed, four cleavage sites by the viral protease were biochemically determined for the PFV isolate. Among these cleavages, one mainly occurs in infected cells and is located at the C-terminus of Gag, leading to a mature product of 68 kDa for PFV (for a more detailed discussion of the cleavages by the viral protease, see the chapter by Flügel and Pfrepper, this volume), a situation already reported for yeast retrotransposons. This cleavage has been observed for FFV (Winkler et al. 1997a) and EFV (Lecellier et al. 2002b). Although the other putative cleavage sites were defined by computer analysis in non-primate PFVs, the direct evidence of cleavage products is still lacking. Moreover, like all PFVs, the canonical zinc finger motifs typically found in other retroviral nucleocapsids are functionally replaced by glycine/arginine-rich regions, the so-called GR boxes (Schliephake and Rethwilm 1994). Three of them were described in primate PFVs: whereas GRII contains a nuclear localisation sequence (NLS) targeting the Gag protein to the nucleus, GRI interacts non-specifically with nucleic acids. GRII-dependent nuclear translocation of Gag was shown to be important for a high proviral load, but its presence is dispensable for infectivity (Meiering et al. 2000). Finally, although GRIII is required for optimal viral infectivity, it has not yet been assigned a specific function. Its absence in FFV and EFV demonstrates its dispensability (Tobaly-Tapiero et al. 2000; Winkler et al. 1997a). Note that FFV Gag, instead of being localised in the nucleus as demonstrated for primate PFV Gag, is found in the cytoplasm of infected cells, in distinct structures close to the perinuclear region (Bodem et al. 1998b). This observation resembles the pericentriolar targeting of PFV Gag during the early

Table 1. Size of predicted PFV gene products and LTR (GenBank accession numbers for the cited viral genomes are U94514 for BFV, Y08851 for FFV, AF201902 for EFV, X54482 for SFV1, U04327 for SFVcpz and Y07725 for PFV)

	BFV	FFV	EFV	SFV1	SFVcpz	PFV
LTR (bp)	1,305	1,355	1,449	1,623	1,769	1,769
Gag (aa)	544	514	559	647	653	648
Pol (aa)	1,220	1,156	1,154	1,158	1,146	1,143
Env (aa)	990	982	987	985	988	989
Tas (aa)	249	209	249	308	300	300
ORF2 (aa)	335	313	329	403	364	356

bp, base pair; aa, amino acid

	⇓		⇓
BFV	GGCTGTG**GT**	.....1720 bp.....	**AG**GCCCAAT
FFV	AGCTTGG**GT**	.....1511 bp.....	**AG**AGAGATC
EFV	AGACTGG**GT**	.....1370 bp.....	**AG**GCCAGTG
PFV	AGGTTTG**GT**	.....1796 bp.....	**AG**GCCTTAA

Fig. 3. Splice sites giving rise to the pol mRNA

phases of infection (Saïb et al. 1997a). In the case of EFV, Gag is found in both, the cytoplasm and in the nucleus. Moreover, in the early stages of infection, EFV Gag migrates to the centrosome using the microtubule network (Lecellier et al. 2002b) as already demonstrated for PFV (Saïb et al. 1997a). As opposed to the *gag* gene, the sizes of the other viral genes are similar for primate and non-primate PFVs (Table 1).

3.2 The *pol* Gene

Although clearly belonging to the retroviral family, PFV presents several specific features relating them to the *Hepadnavirus* family. In particular, they generate a pol mRNA for the synthesis of Pol instead of producing it from a Gag-Pol mRNA as all other known retroviruses by frameshift or translational suppression (Yu et al. 1996; see Fig. 3). This specific mRNA was described for BFV (Holzschu et al. 1998), FFV (Bodem et al. 1996) and, more recently, for EFV (Lecellier et al., 2002b). The protease domain of the Pol polyprotein displays a DSGA (PFV, SFV, BFV), a DSQA (FFV) or a DTGA (EFV) motif as the active centre of the enzyme. The canonical HH-CC zinc finger motif, as well as the catalytic site of the RT enzyme (YVDD), are conserved among all PFVs. Interestingly, PFV Pol harbours two NLS (in the IN and the RT/RNAseH domains) which might act in concert with the Gag NLS in some aspects of the replication cycle such as, for example, nuclear accumulation of viral DNA forms as suggested for HBV (Imrich et al. 2000).

3.3
The *env* Gene

Similar in size and domain structure, the PFV Env glycoprotein presents the classic bipartite organisation leading to the surface (SU) and the transmembrane (TM) mature products after cleavage by a cellular subtilisin-like protease in the late Golgi apparatus. The putative cleavage site of the leader peptide is always located downstream from a large hydrophobic domain and upstream from a charged region (Wang and Mulligan 1999). However, see the chapter by Lindemann and Goepfert in this volume for a closer discussion of this point. PFV envelopes are characterised by two α-helix domains flanking an unusually prolonged central region consisting of β sheets and loops, which distinguishes PFV TM from all other retroviral glycoproteins. Conservation of several cysteine residues as well as the absence of hypervariable regions, which are typical to many other retroviruses, suggest a similar overall structure of the PFV envelopes.

Remarkably, PFVs harbour a short intra-cytoplasmic Env tail, the shortest among the retroviral family. As described in more detail in the chapter by Lindemann and Goepfert (this volume), the cytoplasmic tail of primate PFV TMs is also characterised by the presence of a dilysine retrieval signal for retrograde transport into the endoplasmic reticulum where the virus is thought to bud (Goepfert et al. 1997, 1999). This motif, consisting of two lysines at positions 3 and either 4 or 5 from the carboxy terminus of TM, is absent in BFV and EFV (Fig. 4). Consistent with this observation, EFV mainly buds at the plasma membrane, as observed by electron microscopy (Fig. 1; Tobaly-Tapiero et al. 2000; Lecellier at al. 2002b). More recently, FFV was shown to mainly bud at the plasma membrane, whereas this isolate harbours such a motif in the cytoplasmic tail of Env (M. Löchelt, personal communication). Therefore, it

```
BFV      ...LEKAAGTLFGNVFSILAYAKPVIIGIILIILLLLVIRILRWLAVE--RRRKQE----  990
FFV      ...IKDTAGGIFGTAFSFLGYVKPVLLGFVIIFCIILIIKIIGWLQNT--RKKDQ-----  982
EFV      ...LGKIADGIFGTTFSLLTYAKPVIIGIVVIVLLILIIRILSWLAALGPRRRRREDKGE  986

SFV1     ...LSNTAQGIFGSAVSLLFYAKPILIGIGVILLIALLFKIISWLPGK---PKKN-----  989
SFV3     ...LSNTAQGIFGTTVSILSYAKPILIGIGVILLIAFLFKIVSWLPGK---KKRN-----  982
PFV     ...LSGTAQGIFGTAFSLLGYLKPILIGVGVILLVILIFKIVSWIPTK---KKNQ-----   989
               .    *  .**   * * * **...*  .*  .  . .*. *.      .
                        membrane spanning domain
```

Fig. 4. Protein sequence of the intra-cytoplasmic tail of PFV transmembrane (*TM*) mature product. All PFVs possess an ER retrieval dilysine motif except the bovine and the equine foamy viruses

seems that non-primate PFVs preferentially bud at the plasma membrane.

Whereas the *gag* gene shows a high degree of diversity among PFVs, the relative conservation of *env* sequences compared to what is described for lentiviruses reflects the situation encountered for PTLV group and murine leukaemia viruses (Wang and Mulligan 1999). The reason for the low conservation of gag sequences remains unknown, but this could be related to the distinct mode of production between Gag and Pol, compared to other retroviruses: co-translation of Gag and Pol may impose strict constraints which are absent in the case of PFVs (Wang and Mulligan 1999).

At the 3' end of the *env* gene, a Tas-dependent internal promoter has been described, initially in PFV and further in other primate PFVs, which directs early transcription of the auxiliary genes. By sequence analogy, in particular the presence of a TATA box, a similar transcriptional unit has been deduced in non-primate PFVs. The functionality of this internal promoter has been directly tested in transient transfection assays for FFV (Bodem et al. 1998a) and EFV (Tobaly-Tapiero et al. 2000; Lecellier et al. 2002b) and remains to be evaluated for BFV.

3.4 The Auxiliary Genes

Between the *env* gene and the 3' LTR, two open reading frames (ORF1 and ORF2) were described in non-primate PFVs. Like primate PFVs, ORF1 encodes the viral transactivator Tas which is essential for viral replication and ORF2 encodes the main part of the Bet protein. Remarkably, non-primate PFV transcriptional transactivators are constantly smaller than their primate counterparts (209 for FFV or 249 for BFV and EFV compared to 300 residues of PFV Tas). The main difference lies within the amino-terminal region of Tas, which was reported to be dispensable for the function of the protein (Winkler et al. 1997). All PFV Tas proteins harbour the motif $YXCXXCX_{35-37}R/KH$ of unknown function which is located in an α-helix region predicted to be exposed on the surface of the transactivator (Winkler et al. 1997a). The bipartite NLS described in PFV Tas has been found by computer analysis in EFV Tas, but this basic region seems to be completely absent in BFV and FFV (Tobaly-Tapiero et al. 2000). Interestingly, despite the presence of a consensual NLS in EFV Tas, this transactivator is mainly detected in the cytoplasm of infected cells, resembling the situation encountered for the horse lentiviral Tat transactivator (Lecellier et al. 2002b; Rosin-Arbesfeld et al. 1994). The reason of this difference is not yet determined, but it could be interesting to see whether other transactivators of non-primate PFVs share a similar sub-cellular distribution.

Although structural and enzymatic genes show a relative degree of homology between primate and non-primate PFVs, the homology between known PFV regulatory genes, even among primate and non-primate PFVs, remains low. The highly expressed Bet protein is translated from a multispliced mRNA which fuses an ORF1 exon to ORF2 (Fig. 5). For PFV, the first 88 amino acids of Tas are fused to 394 residues of ORF2, leading to the 60-kDa Bet protein. This protein has been detected in cells infected with FFV and EFV (Winkler et al. 1997; Lecellier et al. 2002a), whereas only the cDNA has been reported for BFV. It is noteworthy that although Bet seems to be dispensable for in vitro replication of PFV, its presence is required for efficient replication of the feline isolate, suggesting that this open reading frame should be included in the design of FFV-derived vectors (Alke et al. 2001; M. Löchelt, unpublished data). Note that the third ORF described in PFV was never observed in non-primate PFVs (Maurer and Flügel 1987).

All PFVs share a complex genomic structure and express their mRNA from two distinct promoters. This leads to a complex pattern of gene expression, which was described for BFV (Renshaw and Casey 1994a,b; Renshaw et al. 1991) and recently for FFV (Bodem et al. 1998a). The model established for PFV, suggesting that the internal promoter is essential for early viral gene expression, also seems to apply for the non-primate PFVs as recently demonstrated for FFV (Bodem et al. 1998a).

A

	⇓		⇓
BFV	GCAGACA**GT**	.....308 bp.....	**AG**TCATCAG
FFV	ACCTCAG**GT**	.....283 bp.....	**AG**TCGGAAA
EFV	TCTGACT**GT**	.....298 bp.....	**AG**TCTTCAG
PFV	GAAATGA**GT**	.....301 bp.....	**AG**TTGCTCA

B

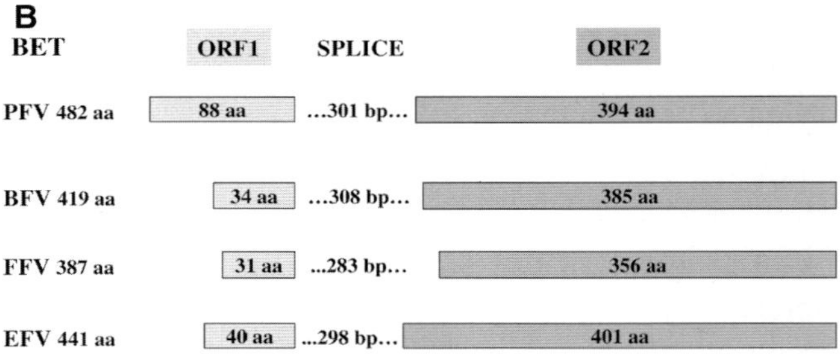

Fig. 5. Splice sites giving rise to the auxiliary Bet protein (*A*) and schematic representation of the corresponding proteins (*B*)

4 Conclusions

Structurally and functionally similar, PFVs are widespread in their natural hosts. Despite the drastic lysis they induce in cell culture, none of these PFV isolates has been shown to induce a pathology or a particular lesion in vivo, suggestive for an efficient retroviral adaptation. Study of the immune response against PFVs should us help to better understand this highly efficient adaptation. Small animal models, such as mice (Schmidt et al. 1997) or rabbits (Saïb et al. 1997b), can be useful in that respect. However, this apparent innocuousness should not outshine a possible transmission of PFVs from primates, and to a lesser extent from non-primates, to humans and the subsequent risk of viral drift. Such cases of zoonotic infection have been reported (see Heneine et al., this volume) but so far humans seem to represent dead-end hosts.

Acknowledgements. The author thanks the members of the foamy virus team of CNRS UPR9051 and critical reading of the manuscript. This work is supported by ARC (#5981) and ECS/SIDACTION.

References

Alke A, Schwantes A, Kido K, Flotenmeyer M, Flügel RM, Löchelt M (2001) The bet gene of feline foamy virus is required for virus replication. Virology 287:310–320

Bodem J, Löchelt M, Winkler I, Flower RP, Delius H, Flügel, RM (1996) Characterization of the spliced pol transcript of feline foamy virus: the splice acceptor site of the pol transcript is located in gag of foamy viruses. J Virol 70:9024–9027

Bodem J, Löchelt M, Delius H, Flügel RM (1998a) Detection of subgenomic cDNAs and mapping of feline foamy virus mRNAs reveals complex patterns of transcription. Virology 244:417–426

Bodem J, Zemba M, Flügel, RM (1998b) Nuclear localization of the functional Bel 1 transactivator but not of the gag proteins of the feline foamy virus. Virology 251:22–27

Chiswell DJ, Pringle CR (1977) Infectious DNA from cells infected with feline syncytium forming virus (spumavirinae). J Gen Virol 36:551–555

Chiswell DJ, Pringle CR (1979a) Feline-syncytium-forming virus: DNA provirus size and structure. J Gen Virol 44:145–152

Chiswell DJ, Pringle CR (1979b) Feline-syncytium-forming virus: identification of a virion associated reverse transcriptase and electron microscopical observations of infected cells. J Gen Virol 43:429–434

Daniels MJ, Golder MC, Jarrett O, MacDonald DW (1999) Feline viruses in wildcats from Scotland. J Wildl Dis 35:121–124

Dutcher RM, Larkin EP, Tumilowicz JJ, Nazerian K, Eusebio CP, Stock ND, Guest GB, Marshak RR (1967) Evidence in support of a virus etiology for bovine leukemia. Cancer 20:851–856

Enders J, Peebles TC (1954) Propagation in tissue cultures of cytopathogenic agents from patients with measles. Proc Soc Exp Biol Med 86:277–286

Fabricant CG, Rich LJ, Gillepsie JH (1969) Felinevirus. IX: isolation of a virus similar to a myxovirus from cats in which urolithiasis was experimentally induced. Cornell Vet 59:667–672

Flanagan M (1992) Isolation of a spumavirus from a sheep. Austr Vet J 69:112–113

Flügel RM (1991) Spumaviruses: a group of complex retroviruses.. J Acquir Immune Defic Syndr 4:739–750

Goepfert PA, Shaw K, Wang G, Bansal A, Edwards BH, Mulligan MJ (1999) An endoplasmic reticulum retrieval signal partitions human foamy virus maturation to intracytoplasmic membranes. J Virol 73:7210–7217

Goepfert PA, Shaw KL, Ritter GD, Mulligan MJ (1997) A sorting motif localizes the foamy virus glycoprotein to the endoplasmic reticulum. J Virol 71:778–784

Helps CR, Harbour DA (1997) Comparison of the complete sequence of feline spumavirus with those of the primate spumaviruses reveals a shorter gag gene. J Gen Virol 78:2549–2564

Holzschu DL, Delaney MA, Renshaw RW, Casey JW (1998) The nucleotide sequence and spliced pol mRNA levels of the nonprimate spumavirus bovine foamy virus. J Virol 72:2177–2182

Hruska JF, Takemoto KK (1975) Biochemical properties of a hamster syncytium-forming virus. J Natl Cancer Inst 54:601–605

Imrich H, Heinkelein M, Herchenroder O, Rethwilm A (2000) Primate foamy virus Pol proteins are imported into the nucleus. J Gen Virol. 81:2941–2947
Jacobs RM, Pollari FL, McNab WB, Jefferson B (1995) A serological survey of bovine syncytial virus in Ontario: associations with bovine leukemia and immunodeficiency-like viruses, production records, and management practices. Can J Vet Res 59:271–278
Johnson RH, de la Rosa J, Abher I, Kertayadnya IG, Entwistle KW, Fordyce G, Holroyd RG (1988) Epidemiological studies of bovine spumavirus. Vet Microbiol 16:25–33
Johnston PB (1974) Studies on simian foamy viruses and syncytium-forming viruses of lower animals. Lab Anim Sci 24:159–166
Kennedy-Stoskopf S, Stoskopf MK, Eckhaus MA, Strandberg JD (1986) Isolation of a retrovirus and a herpesvirus from a captive California sea lion. J Wildl Dis 22:156–164
Kertayadnya IG, Johnson RH, Abher I, Burgess GW (1988) Detection of immunological tolerance to bovine spumavirus (BSV) with evidence for salivary excretion and spread of BSV from the tolerant animal. Vet Microbiol 16:35–39
Lecellier CH, Saïb A (2000) Foamy viruses: between retroviruses and pararetroviruses. Virology 271:1–8
Lecellier CH, Vermeulen W, Bachelerie F, Giron ML, Saïb A (2002a) Intra- and intercellular trafficking of the foamy virus auxiliary Bet protein. J. Virol. 76:3388–3394
Lecellier CH, Neves M, Tobaly-Tapiro J, Giron ML, Saïb A (2002b) Further characterization of the equine foamy virus (EFV) reveals unusual features among foamy viruses. J Virol 76:7220–7227
Malmquist WA, Van Der Maaten MJ, Boothe AD (1969) Isolation, immunodiffusion, immunofluorescence, and electron microscopy of a syncytial virus of lymphosarcomatous and apparently normal cattle. Cancer Res. 29:188–200
Maurer B, Flügel RM (1987) The 3'-orf protein of human immunodeficiency virus 2 shows sequence homology with the *bel3* gene of the human spumaretrovirus. FEBS Lett. 222:286–288
Meiering CD, Comstock KE, Linial ML (2000) Multiple integrations of human foamy virus in persistently infected human erythroleukemia cells. J Virol 74:1718–1726
Pamba R, Jeronimo C, Archambault D (1999) Detection of bovine retrospumavirus by the polymerase chain reaction. J Virol Methods 78:199–208
Pfrepper KI, Löchelt M, Rackwitz HR, Schnölzer M, Heid H, Flügel RM (1999) Molecular characterization of proteolytic processing of the gag proteins of human spumavirus. J Virol 73:7907–7911
Renshaw RW, Casey JW (1994a) Analysis of the 5' long terminal repeat of bovine syncytial virus. Gene 141:221–224
Renshaw RW, Casey JW (1994b) Transcriptional mapping of the 3' end of the bovine syncytial virus genome. J Virol 68:1021–1028
Renshaw RW, Gonda MA, Casey JW (1991) Structure and transcriptional status of bovine syncytial virus in cytopathic infections. Gene 105:179–184
Riggs JL, Oshiro LS, Taylor DO, Lennette EH (1969) Syncytium-forming agent isolated from domestic cats. Nature 222:1190–1191

Rosin-Arbesfeld R, Mashiah P, Willbold D, Rosch P, Tronick SR, Yaniv A, Gazit A (1994) Biological activity and intracellular location of the Tat protein of equine infectious anemia virus. Gene 150:307–311

Saïb A, Périès J, de Thé H (1995) Recent insights into the biology of the human foamy virus. Trends Microbiol 3:173–178.

Saïb A, Puvion-Dutilleul F, Schmid M, Périès J, de Thé H (1997a) Nuclear targeting of incoming human foamy virus gag proteins involves a centriolar step. J Virol 71:1155–1161

Saïb A, Neves M, Giron ML, Guillemin MC, Valla J, Périès G, Canivet M (1997b) Long term persistent infection of rabbits by the human foamy virus. Virology 228:263–268

Schliephake AW, Rethwilm A (1994) Nuclear localization of foamy virus gag precursor protein. J Virol 68:4946–4954.

Schmidt M, Niewiesk S, Heeney J, Aguzzi A, Rethwilm A (1997) Mouse model to study the replication of primate foamy viruses. J Gen Virol 78:1929–1933

Tobaly-Tapiero J, Bittoun P, Neves M, Guillemin MC, Lecellier CH, Puvion-Dutilleul F, Gicquel B, Zientara S, Giron ML, de Thé H, Saïb A (2000) Isolation and characterization of an equine foamy virus. J Virol 74:4064–4073

Wang G, Mulligan MJ (1999) Comparative sequence analysis and predictions for the envelope glycoproteins of foamy viruses. J Gen Virol 80:245–254

Winkler I, Bodem J, Haas L, Zemba M, Delius H, Flower R, Flügel RM, Löchelt M (1997a) Characterization of the genome of feline foamy virus and its proteins shows distinct features different from those of primate spumaviruses. J Virol 71:6727–6741

Winkler IG, Flügel RM, Löchelt M, Flower RLP (1998) Detection and molecular characterisation of feline foamy virus serotypes in naturally infected cats. Virology 247:144–151

Winkler IG, Löchelt M, Flower RLP (1999) Epidemiology of feline foamy virus and feline immunodeficiency virus infections in domestic and feral cats: a seroepidemiological study. J Clin Microbiol 37:2848–2851

Winkler IG, Löchelt M, Levesque JP, Bodem J, Flügel RM, Flower RLP (1997b) A rapid streptavidin-capture ELISA specific for the detection of antibodies to feline foamy virus. J.Immunol Methods 207:69–77

Yu SF, Baldwin, DN, Gwynn SR, Yendapalli S, Linial ML (1996) Human foamy virus replication: a pathway distinct from that of retroviruses and hepadnaviruses. Science 271:1579–1582

Zenger E, Brown WC, Song W, Wolf AM, Pedersen NC, Longnecker M, Li J, Collisson EW (1993) Evaluation of cofactor effect of feline syncytium-forming virus on feline immunodeficiency virus infection. Am J Vet Res 1993: 13–718

Subject Index

Current Topics in Microbiology and Immunology

Volumes published since 1989 (and still available)

Vol. 233/II: **Tyler, Kenneth L.; Oldstone, Michael B. A. (Eds.):** Reoviruses II. 1998. 45 figs. XVI, 187 pp. ISBN 3-540-63947-0

Vol. 234: **Frankel, Arthur E. (Ed.):** Clinical Applications of Immunotoxins. 1999. 16 figs. IX, 122 pp. ISBN 3-540-64097-5

Vol. 235: **Klenk, Hans-Dieter (Ed.):** Marburg and Ebola Viruses. 1999. 34 figs. XI, 225 pp. ISBN 3-540-64729-5

Vol. 236: **Kraehenbuhl, Jean-Pierre; Neutra, Marian R. (Eds.):** Defense of Mucosal Surfaces: Pathogenesis, Immunity and Vaccines. 1999. 30 figs. IX, 296 pp. ISBN 3-540-64730-9

Vol. 237: **Claesson-Welsh, Lena (Ed.):** Vascular Growth Factors and Angiogenesis. 1999. 36 figs. X, 189 pp. ISBN 3-540-64731-7

Vol. 238: **Coffman, Robert L.; Romagnani, Sergio (Eds.):** Redirection of Th1 and Th2 Responses. 1999. 6 figs. IX, 148 pp. ISBN 3-540-65048-2

Vol. 239: **Vogt, Peter K.; Jackson, Andrew O. (Eds.):** Satellites and Defective Viral RNAs. 1999. 39 figs. XVI, 179 pp. ISBN 3-540-65049-0

Vol. 240: **Hammond, John; McGarvey, Peter; Yusibov, Vidadi (Eds.):** Plant Biotechnology. 1999. 12 figs. XII, 196 pp. ISBN 3-540-65104-7

Vol. 241: **Westblom, Tore U.; Czinn, Steven J.; Nedrud, John G. (Eds.):** Gastroduodenal Disease and Helicobacter pylori. 1999. 35 figs. XI, 313 pp. ISBN 3-540-65084-9

Vol. 242: **Hagedorn, Curt H.; Rice, Charles M. (Eds.):** The Hepatitis C Viruses. 2000. 47 figs. IX, 379 pp. ISBN 3-540-65358-9

Vol. 243: **Famulok, Michael; Winnacker, Ernst-L.; Wong, Chi-Huey (Eds.):** Combinatorial Chemistry in Biology. 1999. 48 figs. IX, 189 pp. ISBN 3-540-65704-5

Vol. 244: **Daëron, Marc; Vivier, Eric (Eds.):** Immunoreceptor Tyrosine-Based Inhibition Motifs. 1999. 20 figs. VIII, 179 pp. ISBN 3-540-65789-4

Vol. 245/I: **Justement, Louis B.; Siminovitch, Katherine A. (Eds.):** Signal Transduction and the Coordination of B Lymphocyte Development and Function I. 2000. 22 figs. XVI, 274 pp. ISBN 3-540-66002-X

Vol. 245/II: **Justement, Louis B.; Siminovitch, Katherine A. (Eds.):** Signal Transduction on the Coordination of B Lymphocyte Development and Function II. 2000. 13 figs. XV, 172 pp. ISBN 3-540-66003-8

Vol. 246: **Melchers, Fritz; Potter, Michael (Eds.):** Mechanisms of B Cell Neoplasia 1998. 1999. 111 figs. XXIX, 415 pp. ISBN 3-540-65759-2

Vol. 247: **Wagner, Hermann (Ed.):** Immunobiology of Bacterial CpG-DNA. 2000. 34 figs. IX, 246 pp. ISBN 3-540-66400-9

Vol. 248: **du Pasquier, Louis; Litman, Gary W. (Eds.):** Origin and Evolution of the Vertebrate Immune System. 2000. 81 figs. IX, 324 pp. ISBN 3-540-66414-9

Vol. 249: **Jones, Peter A.; Vogt, Peter K. (Eds.):** DNA Methylation and Cancer. 2000. 16 figs. IX, 169 pp. ISBN 3-540-66608-7

Vol. 250: **Aktories, Klaus; Wilkins, Tracy, D. (Eds.):** Clostridium difficile. 2000. 20 figs. IX, 143 pp. ISBN 3-540-67291-5

Vol. 251: **Melchers, Fritz (Ed.):** Lymphoid Organogenesis. 2000. 62 figs. XII, 215 pp. ISBN 3-540-67569-8

Vol. 252: **Potter, Michael; Melchers, Fritz (Eds.):** B1 Lymphocytes in B Cell Neoplasia. 2000. XIII, 326 pp. ISBN 3-540-67567-1

Vol. 253: **Gosztonyi, Georg (Ed.):** The Mechanisms of Neuronal Damage in Virus Infections of the Nervous System. 2001. approx. XVI, 270 pp. ISBN 3-540-67617-1

Vol. 254: **Privalsky, Martin L. (Ed.):** Transcriptional Corepressors. 2001. 25 figs. XIV, 190 pp. ISBN 3-540-67569-8

Vol. 255: **Hirai, Kanji (Ed.):** Marek's Disease. 2001. 22 figs. XII, 294 pp. ISBN 3-540-67798-4

Vol. 256: **Schmaljohn, Connie S.; Nichol, Stuart T. (Eds.):** Hantaviruses. 2001, 24 figs. XI, 196 pp. ISBN 3-540-41045-7

Vol. 257: **van der Goot, Gisou (Ed.):** Pore-Forming Toxins, 2001. 19 figs. IX, 166 pp. ISBN 3-540-41386-3

Vol. 258: **Takada, Kenzo (Ed.):** Epstein-Barr Virus and Human Cancer. 2001. 38 figs. IX, 233 pp. ISBN 3-540-41506-8

Vol. 259: **Hauber, Joachim, Vogt, Peter K. (Eds.):** Nuclear Export of Viral RNAs. 2001. 19 figs. IX, 131 pp. ISBN 3-540-41278-6

Vol. 260: **Burton, Didier R. (Ed.):** Antibodies in Viral Infection. 2001. 51 figs. IX, 309 pp. ISBN 3-540-41611-0

Vol. 261: **Trono, Didier (Ed.):** Lentiviral Vectors. 2002. 32 figs. X, 258 pp. ISBN 3-540-42190-4

Vol. 262: **Oldstone, Michael B.A. (Ed.):** Arenaviruses I. 2002, 30 figs. XVIII, 197 pp. ISBN 3-540-42244-7

Vol. 263: **Oldstone, Michael B. A. (Ed.):** Arenaviruses II. 2002, 49 figs. XVIII, 268 pp. ISBN 3-540-42705-8

Vol. 264/I: **Hacker, Jörg; Kaper, James B. (Eds.):** Pathogenicity Islands and the Evolution of Microbes. 2002. 34 figs. XVIII, 232 pp. ISBN 3-540-42681-7

Vol. 264/II: **Hacker, Jörg; Kaper, James B. (Eds.):** Pathogenicity Islands and the Evolution of Microbes. 2002. 24 figs. XVIII, 228 pp. ISBN 3-540-42682-5

Vol. 265: **Dietzschold, Bernhard; Richt, Jürgen A. (Eds.):** Protective and Pathological Immune Responses in the CNS. 2002. 21 figs. X, 278 pp. ISBN 3-540-42668-X

Vol. 266: **Cooper, Koproski (Eds.):** The Interface Between Innate and Acquired Immunity, 2002, 15 figs. XIV, 116 pp. ISBN 3-540-42894-1

Vol. 267: **Mackenzie, John S.; Barrett, Alan D. T.; Deubel, Vincent (Eds.):** Japanese Encephalitis and West Nile Viruses. 2002. 66 figs. X, 418 pp. ISBN 3-540-42783-X

Vol. 268: **Zwickl, Peter; Baumeister, Wolfgang (Eds.):** The Proteasome-Ubiquitin Protein Degradation Pathway. 2002, 17 figs. X, 213 pp. ISBN 3-540-43096-2

Vol. 269: **Koszinowski, Ulrich H.; Hengel, Hartmut (Eds.):** Viral Proteins Counteracting Host Defenses. 2002, 47 figs. XII, 325 pp. ISBN 3-540-43261-2

Vol. 270: **Beutler, Bruce; Wagner, Hermann (Eds.):** Toll-Like Receptor Family Members and Their Ligands. 2002, 31 figs. X, 192 pp. ISBN 3-540-43560-3

Vol. 271: **Koehler, Theresa M. (Ed.):** Anthrax. 2002, 14 figs. X, 169 pp. ISBN 3-540-43497-6

Vol. 272: **Doerfler, Walter; Böhm, Petra (Eds.):** Adenoviruses: Model and Vectors in Virus Host Interactions. Virion and Structure, Viral Replication, Host Cell Interactions. 2003, 63 figs., approx. 280 pp. ISBN 3-540-00154-9

Vol. 273: **Doerfler, Walter; Böhm, Petra (Eds.):** Adenoviruses: Model and Vectors in Virus Host Interactions. Immune System, Oncogenesis, Gene Therapy. 2003, 33 figs., approx. 280 pp. ISBN 3-540-06851-1

Vol. 274: **Workman, J. L. (Ed.):** Protein Complexes that Modify Chromatin. 2003, 38 figs., XII, 296 pp. ISBN 3-540-44208-1

Vol. 275: **Fan, Hung (Ed.):** Jaagsiekte Sheep Retrovirus and Lung Cancer. 2003, 63 figs., XII, 252 pp. ISBN 3-540-44096-3

Vol. 276: **Steinkasserer, A. (Ed.):** Dendritic Cells and Virus Infection. 2003, 24 figs., X, 296 pp. ISBN 3-540-44290-1

Printing (Computer to Plate): Saladruck Berlin
Binding: Stürtz AG, Würzburg